SNÖ

www.penguin.co.uk

S

Sverker Sörlin

N

A History

Ö

**Translated from Swedish
by Elizabeth DeNoma**

doubleday

TRANSWORLD PUBLISHERS

UK | USA | Canada | Ireland | Australia
India | New Zealand | South Africa

Transworld is part of the Penguin Random House group of companies whose
addresses can be found at global.penguinrandomhouse.com.

Penguin Random House UK, One Embassy Gardens, 8 Viaduct Gardens,
London SW11 7BW

penguin.co.uk

Penguin
Random House
UK

First published in Great Britain in 2025 by Doubleday
an imprint of Transworld Publishers

001

Typeset in 11.1/15.2pt Calluna by Six Red Marbles UK, Thetford, Norfolk
Printed and bound in Great Britain by Clays Ltd, Elcograf S.p.A.

The authorized representative in the EEA is Penguin Random House Ireland,
Morrison Chambers, 32 Nassau Street, Dublin D02 YH68.

A CIP catalogue record for this book is available from the British Library

ISBNs
9781529947878 (hb)
9781529947885 (tpb)

Penguin Random House is committed to a sustainable
future for our business, our readers and our planet. This book is
made from Forest Stewardship Council® certified paper.

Hast thou entered into the treasures of the snow?
Or hast thou seen the treasures of the hail,
Which I have reserved against the time of trouble, against the day of
battle and war?

Hath the rain a father?
Or who hath begotten the drops of dew?
Out of whose womb came the ice?
And the hoary frost of heaven, who hath gendered it?

Job 38: 22–23, 28–29

Contents

Contents

Introduction: Snow Angel

*

SNOW ANGEL.

Pay attention to these words. See the image in front of you. An angel in the snow. A white figure with wings spread wide.

I'm lying on my back with my eyes momentarily closed. I move my arms upward and downward. Soon I've created an improved version of myself.

I have always thought that a book about snow should start with those words. Everyone living at the northern latitudes knows what a snow angel is. Many people have themselves made them. Do angels exist? Snow angels exist. That is something certain.

Snow angels exist for a moment in time. Until the new snow falls, covering them up; until some angel-hater destroys them; until they melt away. An angel made of snow is a beautiful image of hope and impermanence, created out of a human desire for a different, better, state of things. The kind of desire which nowadays might sometimes be called childish.

A snow angel is the opposite of Jewish-German philosopher Walter Benjamin's 'Angel of History' – the angel with its eyes wide open in terror, its wings stretching towards the sky but paralysed by fear, its rigid, anguished body moving backwards into the future, hurled in that direction by a storm emanating out of Paradise. The horror never loosens its grip on the Angel of History, because the angel sees before it all the suffering and evil that history has produced. All the terrible things people have done to each other and to Creation. Everything that the rest of us don't see when we have to turn our backs on the past, looking ahead, to see where to place our feet, perhaps filled with hope, confidence and the positive thought that what lies ahead of us may be more beautiful.

The angel sees what we leave behind and dare not look back upon – for fear of perishing.

Benjamin, who later saw his own books destroyed by the Nazis in their book-burnings, came up with his idea in 1921, when he'd bought a painting of just such an angel by the Swiss artist Paul Klee. The most awful thing about the Angel of History is that Benjamin's interpretation is correct. That's what history is. And that's what the future will also be, at least to some extent. To reduce some part of that evil and suffering, we humans must take some responsibility.

This is how I see the role of the snow angel in our lives: as mankind's way of exorcizing the future and strengthening us in our mission to improve ourselves. With my eyes on the constellation of stars above me, with moral responsibility within me and the whole of the earth beneath me, covered with this fragile, soft white blanket of impermanence, I ward off the forces of evil with my simple angelic gesture, demonstrating with my fumbling wings that the earth can transform into Paradise. The way that we have to believe that Sisyphus, eternally rolling his boulder uphill, is happy. That it's meant to be, even if we know that's not exactly what it's going to be like. I want these two very different angels – of history and of snow – to at least be able to talk to each other, and I think that's why I've always wanted to write this book.

Snow is hope – renewed annually, with new snow replacing that which fell last year and which has already disappeared. In literature, they say that snow is a symbol of transformation. When there's snowfall in poetry and novels, it's often to demonstrate that the characters in the text are turning inwards in reflection. They become receptive to forces or messages that can help them reach another, better, state of being.

Snow is also history, having been a companion to people's lives in many parts of the world, especially in the northern hemisphere, and even more so for those of us living in the far north. It's about time that the snow angel and the Angel of History started conversing.

What's happening to snow is something that the Angel of History is seeing more and more of as we humans make our way through the world. Where once there was a real winter with snow, now bare ground makes an appearance during seasons that have lost their rhythm. Snow is in retreat, as part of the great historical drama now known as climate change. It's not the first climate change that history's all-seeing angel has witnessed. But it's the first that human beings have themselves brought about on a planetary scale. The first climate change that is, in the deepest sense of the word, *historical*, that is to say, part of the history of human actions.

When was it that the first snow fell on our planet? The very question itself previously seemed comically fantastic. Now there's an answer: 2.5 billion years ago. In 2018, a group of geologists published an article in the journal *Nature*, presenting a theory for how large continental land masses first emerged out of the seas, and how Earth subsequently started reflecting light from the sun back into space. The planet's temperature dropped, causing runaway glaciation that eventually left it in a state known as 'Snowball Earth'. Ilya Bindeman, a professor of geology at the University of Oregon, who led the research team, said in an interview, 'That's when the earth got its first snowfall.'

The world's oldest ski is 5,400 years old. More recent than the first snowfall, but still long before anyone uttered the word 'Sweden'. It was found in 1923 near the village of Kalvträsk, between Skellefteå and Lycksele in what is present-day Västerbotten, the second-northernmost county of Sweden. Another ski, almost as old, was later found in Hoting, not far from the Ångermanland border, and less than forty kilometres from where I grew up. In sparse, rugged Sweden, people in the countryside were reluctant to carry their loads across the bare ground, in terrain without roads. Winter was better; anything heavy and unwieldy could be towed home over the snow – things like timber or firewood from the forest, or hay that had been cut and hung on hayracks or put into barns along the river banks. Skis are faster than snowshoes, especially in areas where temperatures sometimes hover around zero, causing the snow to melt and freeze. On skis you could pursue moose and bears, which had a hard time in deep snow. Snow meant freedom and protein.

Before steamboats or railways existed, merchants in the north transported goods by horse and sled along the Gulf of

Bothnia coast. Our vast country was held together by snow and ice. Snow was the basic prerequisite.

Growing up as a little boy in the snowy interior of Väster-botten, and in southernmost Lapland, in northern Sweden, snow was never something 1 associated with 'history'. History, we were taught at school, was what had happened in other places. Almost always further south, where Sweden's kings had fought wars in continental Europe, and where popes and parliaments made important decisions. These decisions were certainly relevant to us and affected our lives. But they didn't affect the basic conditions of our existence, which when it came right down to it, were more influenced by the seasons, and by the climate.

Snow was a part of that. It was indeed the underlying foundation of our lives, if one that wasn't always appreciated. Loggers cut down large timber for export, making Sweden rich. They did their work under mighty tree branches that were usually weighed down by thick layers of snow – *opplegan*, as it was called in our dialect. Snow that fell on those loggers had a strange way of making it through the collars of their shirts, chilling their necks, melting and running down their backs and dampening their thick underwear with snow-melt. That was before the term 'work environment' came into use. But the snow was, at its heart, something practical, and without it the development of the forestry industry would have been limited. It was on the snow that timber was transported to lakes, streams and rivers. And it was on the snow that the loggers made their way out to the clearings, mostly on skis. Some of them became skilful racers, in the generations that dominated cross-country skiing from the Second World War until the 1970s.

I grew up on snow, too. There was snow for more than half the year. That's not a fact I've pulled out of thin air or even from memory; I've consulted records from the Swedish Meteorological and Hydrological Institute (SMHI). The number of snow days per year in northern Norrland, which includes Västerbotten and Lapland, averaged between 180 and just over 220 throughout my childhood and youth. The peak was the winter of 1968–69, with snow cover lasting nearly eight months even in places outside the mountain range. Snow was the normal condition, as was the cold. The snow was such an obvious thing that there was no reason to waste time on it. No one thought it had anything to do with history. Or that it could be connected to politics. My grandfather was a politician, sitting in parliament, the county council and the municipal governing body. But when he talked about politics, as he sometimes did at home with us, it was never about snow. Snow just existed, like something eternal that had to be shovelled from All Saints' Day until Walpurgis Night.

It's just that I loved it. Others sighed; I was happy. My joy came spontaneously, irresistibly. I didn't know why. That's just the way it was. I wasn't at all alone in this feeling. Most of us liked the snow when we were little. People only started complaining about it as adults. I realize now that what I sensed in myself even then, without having words for it, was that snow can be understood, at its deepest sense, as history. It's also physics, chemistry, meteorology, anthropology, economics, technology and geography. Yes, poetry, art, music, dance and sculpture, too. And very much politics.

It's important right now to see that snow is historical. It appears within time. And as the kind of element that compels our attention, snow is growing – in our consciousness, it's getting bigger. But out there, in the environment, it's getting smaller. It's disappearing. This is our own doing. Snow isn't simply a mysterious gift from the heavens. It's increasingly becoming recognized as a part of history. At the same time, in many places it's already history, in the sense that it no longer exists.

I feel this within me. Snow is inside me and it's painful when it disappears. I have images from my childhood in the snow that I carry with me in my life. Only recently have I been able to make sense of them.

Snow – just a lone, single word for an infinite variety of individual expressions of water, frozen in air. We sometimes hear the word from the teacher standing behind the lectern, but the teacher's way of describing snow doesn't impress us very much. As a boy, I am told that in Greenland there are a hundred words for snow. I wonder to myself if that is true. Maybe it is – after all, snow *is* marvellous. But still . . . is that possible? Isn't it just another one of those exaggerations that the world seems to be full of? I have realized early on that much of what we hear is simplification; received wisdom that isn't always true.

Many years later, I was to read the author and ethnologist Yngve Ryd's book on words for snow in the Lule Sámi language. It contains more than three hundred words, passed on to Ryd by the reindeer herder Johan Rassa and his wife Ibb-Anna from Jokkmokk. The book is unique – it has no real equivalent anywhere else in the world. In the 1940s, the national dialect archive located in Uppsala sent out a list of questions about snow, but the answers are 'succinct and sparse', says Ryd. From out of his and the Rassas' book, however, a whole world emerges, like a powerful symphony of knowledge, experience and poetry about a culture and a landscape occupied by animals and people and myths. And this is still only the beginning of the world of snow. Snow is a state of being – an internal one as well as an external one. One for the place and one for the planet.

I like the snow. In almost any form it takes. Adults tend to be dubious about it. A child, on the other hand, rushes out, eagerly grasps the white substance and immediately begins to shape it. As if they had caught hold of something that could become ... a new world. That's the way I am. Quite naturally and without knowing why. And when I think back to my childhood self now, I realize that even then, without knowing it, I perceived the snow as historical. Something that was part of being human in the world, and something responsible for shaping that experience.

But I didn't know what we now know. That snow is no longer a given, the way that I thought it was. It's at risk. It challenges us. It's our conscience.

Snow is on its way out. At an uneven pace, but inexorably. Now, more than ever, I really want to understand its qualities. This feels absolutely essential. There's no other substance as important to us about which we know so little. Snow is a transitional substance. Neither stable nor certain. A liminal form. At the same time, it's ordinary water, but in an ambiguous position somewhere between free molecules in the air and the hard, stable ice that can last for millions of years.

It's snowing – this airy substance that appears only from time to time, under very specific, and actually quite infrequent, conditions, which happen to be prevalent to an unusual extent in the Nordic countries. What I've always considered a given is, on the contrary, something rare. And it's becoming more and more rare. That's because we humans are warming the planet. And this is exactly what compels us to turn our attention to snow right now. To ask questions of it, to want to know more about it. Like how it's come into being, what it is, what it's no longer able to be. We can use it to learn more about the dangers of our current moment. Snow can help us to see our self-inflicted misfortune.

Those of us living among the forests and mountains in the Nordic countries have a special responsibility for this task. We know the peculiarities of snow; its strangeness. And we're learning new things about it all the time. Like the fact that monkeys also make snowballs, but they don't have snowball fights. And that snow has politics. Contrary to what we thought – that it just fell and fell and would always fall without caring what humans did. But it turns out that it does care.

Perhaps the strangest thing is that the qualities of snow also reside within us. Those of us who live and have lived with snow have, I think, internalized some of its qualities as part of ourselves. Whether we love the snow or not. Perhaps with horror at the cold of it and the disruption snow sometimes causes; or, as in my case, with deep affection and a warm trust in the snow as our wise ally.

I've always had this trust. Alongside a strange happiness that I've been able to retain through the intervention of some unknown angel. A fixed point in my mind that I can return to when I find it hard to breathe. Snow is the most important element in the reliable wheel of the seasons.

We need all of the seasons, because they're not the same; they have different missions. Autumn is the second-best; it is the anticipation of winter, which is the best season of all. The leaves fall early in inner Västerbotten – a couple of hours from Ådalen's timber industry, on the Gulf of Bothnia in the east, and a few hours from Trøndelag in Norway, by the Atlantic in the west.

And then, one morning, from my boyhood bedroom: I see the light outside, before I even open the slatted blinds. Snow! The miracle of whiteness is shining through the slats. The ground that was dark yesterday is bright and radiant this autumn morning. No human being has been involved. Not even my grandfather, who usually controls so much.

Snowfall is an *event*. Something that comes to me. I'm a part of something, through forces that I have no control over. How fortunate that I, of all people, should happen to be in a place in the world where it snows! I love it because it's a gift. It is the very order of the world; we're all so small that we cannot compete with either the snow or the sky. They're bigger than we are. And maybe that's why I am so happy, and take such childish delight in being alive.

A silent collecting of snow. Eventually this becomes my project. Voices, images, stories. The more I collect, the more convinced I am that a deeper sense of snow is needed. It was ice and snow that made Sweden – we live in a winter wonderland. Why don't we all care more? Why don't we mourn the fact that the snow is melting?

I want to awaken the snow angel within us.

Diamonds
and Birds' Wings

*

I'VE LEARNED TO WALK. My knitted ski helmet is on my head. Mum is in front of me, smiling, sparkling. I'm standing on my first skis. They're blue. No span, no glide.

When I'm going to sleep, my mother reads aloud from Elsa Beskow's *Ollie's Ski Trip*, or the books about Moominvalley. Snow belongs to the world of fairy tales. The books contain a truth that I make my own. Mum's voice brings it to me. She doesn't just read the words. She also lets me understand that this is something she believes in, and something that she wants me to believe.

Mum turns her face towards me. Rubs snow against my cheeks. I suppose I laugh. Her brown eyes and white teeth gleam. She was twenty-two when I was born; I will later realize that she's beautiful. My mother's voice becomes my *Weltanschauung*.

First of all there's the snow itself. Snow is the finest droplets of water, frozen in air, helped by dust.

Next come the words. The Swedish word for 'snow', *snö*, comes from the Old Swedish *sniö* or *snior*, which in turn comes from the Old Norse *snær* (which is also a first name

I

in Icelandic). That in turn comes from the proto-Germanic *snaiwa* (from which also come the English *snow*, the German *Schnee* and the Dutch *sneeuw*). The Indo-European root of all these words is *(s)neyg* (or *snoyg*, *snoigu-ho*), from which the French *neige* and the Italian *neve* have also come about. Along the way, the Latin word *nix*, meaning snow, came via the Proto-Italic (Etruscan) *sniks*, with the root *y*, which gave the prefix *niv-*, which is found in the word nival, as found in the expression 'the subnival space' – the space between the ground and the snow – or in the Sierra Nevada ('snow mountains').

Snow is a rich wordsmith. But perhaps it could just as easily be said that the reverse is true: snow emerges from these words. As the Gospel of John says, 'In the beginning was the Word'. Without words, snow wouldn't be so rich in meaning. Without words, we literally would not see the snow for all its whiteness. When physicist Henri Bader and his team of researchers in Davos, Switzerland, published a book on what they called the 'metamorphism' of snow in the fateful year of 1939, he began the book with a short three-page chapter about the word itself, *Der Begriff 'Schnee'* ('The Concept of "Snow"'). The gist of this was that when it comes to snow, nothing is stable. Snow is metamorphosis; its transformation never ceases. Bader's approach is reminiscent of how Luis Buñuel began his 1929 film *An Andalusian Dog* – a razor blade through a woman's eye (in reality a reindeer's eye). It was something shocking! To make the reader receptive.

To be more precise, snow forms from water vapour in the upper layers of air that cools as the air moves upwards. That water vapour then condenses into fine, flat ice crystals with a six-armed structure. In these conditions this crystallization process needs something material for it to get started: particles, dust, a few molecules.

It's the water molecule, with one oxygen atom and two hydrogen atoms, that readily forms hexagonal crystals. Snow crystals and snowflakes aren't droplets of water that have frozen, but are made up of water molecules that change directly from vapour to solid form. Their structure can vary greatly depending on the temperature, humidity and other atmospheric conditions. If it's really cold, between -12° and -16° Celsius, the ice crystal grows easily and forms the classic, symmetrical snow star with six arms or more, in some exceptional instances twice as many. When it's colder, the snowflake becomes rounder. If it's not particularly cold, the snowflake takes on a simpler shape.

Snowflakes fall at different rates depending on their size, density, temperature and the wind. But usually they fall quite slowly, around five kilometres per hour, about normal walking speed. That humanizes snow. It also moves the same way we do, crookedly and winding, floating and searching, turning sharply.

A snowflake is a collection of many snow stars – dozens, sometimes more than a hundred. Really big snowflakes form when the temperature is higher and the snowflakes are moist on the surface and stick together, which happens as the snowflakes are falling. When there's no wind and it's not too cold, snowflakes can become the size of butterflies. In Lule Sámi they are called sparrow halves, *tsihtsebelaga*. The word 'half' seems to imply two halves that want something from each other. In Swedish, we used to call large snow-flakes *lapphandskar* – that is, Lapp gloves (or Sámi gloves). My mother Gudrun would say, reverently, 'It's snowing Lapp gloves!' Then we children would come up to the window and look out and be filled with that reverence ourselves. *Lapp-handskar* were something lovely, and I would watch with my own eyes as the large white flakes quietly and mysteriously floated to the ground, like in a film.

The expression has been around for a long time, at least a couple of hundred years. Fredrik Svenonius, the state geologist and fierce advocate for the Swedish region of Norrbotten, who was also a keen photographer, took a picture of a snowy landscape with his camera on 20 February 1899 and called it 'It's snowing Lapp gloves'. Petrus Læstadius, a missionary to Lapland, noted in his diary in the early 1830s: 'In Norrland they still say that it's snowing "Lapp gloves", when the snow falls in large flakes.' From this, we can conclude that the expression has been in use since long before that.

Such 'Lapp glove' snowflakes can be huge. The largest known snowflake is said to have been found at Fort Keogh in Montana, USA, on 28 January 1887. During a snowfall, a man named Matt Coleman is said to have come across a snowflake 38 centimetres wide and 20 centimetres thick. In Siberia in 1971, snowflakes 'the size of A4 sheets' were observed by several witnesses. In March 2024, snowflakes 'the size of snuff-boxes' were reported in the northern Swedish city of Umeå.

Snow can fall even when the temperature is above freezing, especially when the humidity is low, as is often the case in late winter and early spring. It can be 7°C outside and you might have caught a glimpse of the first blooms of spring – then suddenly your view is obscured by a swirling flurry of snow. But it usually passes quickly. When snow forms in stable, low-level clouds – typically stratus or stratocumulus – it usually falls as flat, oblong, brittle ice particles rarely more than a few millimetres long, known as granular snow, which can be irritating to the eyes when it's windy. So-called ice needles are even smaller. They fall from clear skies and consist of tiny, thin ice crystals that appear to float in the air. These are also called 'diamond dust'. When lit by the sun, they create a shimmering effect reminiscent of gilded altarpieces. If there is a city somewhere above the clouds, the way there is probably through such dust clouds of light-transmitting ice.

Hail, too, counts as snow, and there are several different kinds. One type is called snow hail or graupel. The particles in graupel are similar to granular snow but larger, up to five millimetres in diameter. Snow hail bounces and often breaks on contact with the ground; sleet, meanwhile, comes in huge waves and doesn't break apart. It can create rivers where it falls.

Hail falls from huge clouds and is sometimes accompanied by thunder, regardless of the season. It can fall at speeds of 150 to 200 kilometres per hour and can reach the size of grapes, plums – or even larger. Ice hail, the sky's heaviest artillery, can dent cars and pierce shingled roofs; it can probably even fell an ox. The largest hailstone recorded under scientifically controlled conditions fell during a storm in Vivian, South Dakota, on 23 July 2010. It was 20.3 centimetres in diameter and weighed 0.88 kilograms. But there are written reports of even larger hailstones. One that landed in China in 1902 is said to have weighed four kilograms. One of the first Westerners to visit Tibet, in the 1840s, the French missionary Évariste Régis Huc witnessed hailstones there weighing five and a half kilograms, one of which was the size of a millstone and took more than three days to melt in the blazing sun.

Compressed snow eventually becomes very hard, transforming over time into ice and, across widespread mountainous areas, into glaciers with properties similar to minerals. Snow itself can be classed as a mineral. It fulfils all the criteria: it's homogeneous, non-organic, with a defined chemical composition (H_2O) and a definite atomic structure. Like other minerals, snow occurs naturally, though mostly at temperatures below 0° Celsius.

Almost all snowfall – around 98 per cent of it – occurs in the northern hemisphere. In a typical winter, snow covers up to 40 per cent of the northern hemisphere's land surface – more than 40 million square kilometres. Of course, the area covered varies from year to year, but the long-term trend is that it is shrinking. Snow is mainly found in North America, Greenland, Europe and Russia, but snow also falls in the Andes and Asia, as well as in New Zealand, south-eastern Australia, Patagonia and Antarctica. In Africa, snow falls in mountainous regions in the north, east and south of the continent. The Asian and Pacific archipelago, West and Central Africa, and central and western Australia, are areas where it virtually never snows. This is also the case in some desert regions of the Andes and in the driest areas of East Antarctica – the McMurdo Dry Valleys – which scientists believe have had their extreme climate for millions of years and consequently have no ice sheet, unlike most of the Antarctic continent.

In Sweden, most of the snow falls in the north, with less in the south. In Skåne, which is in the south, an average of almost 10 per cent of annual precipitation came in the form

of snow during 1960–1990. In Lapland, the corresponding figure was 40 per cent. That difference has probably
decreased slightly since 1990, with the increasingly warm
winters. The greatest proportion of annual precipitation
is in the mountainous regions. At high altitudes, in Sarek
National Park, the Kebnekaise massif and other high mountain areas, the proportion can be 70 per cent or more. It's
also in the mountains that the greatest depths of snow in
Sweden have been measured.

The deepest was at Kopparåsen, near the Norwegian
border, where on 28 February 1926 it is said to have measured
327 centimetres. The snow at nearby weather monitoring
sites was nothing like as deep at the time, so even the field
staff admitted there was some uncertainty surrounding
this measurement. The Swedish Meteorological and Hydrological Institute (SMHI) states, 'Unfortunately, the last day
of February was the last day when observations were made
from the station and we don't know what happened later to
the snow depth or the person who observed it.'

In the village of Ulvoberg, in the Västerbotten region,
one metre of fresh snow fell in June of 1932. A contemporary
source says, 'The snowfall and subsequent large amounts of
meltwater caused a shortage of fodder in several places, and
emergency aid had to be sent to the worst affected areas.'

In December 1998, winds from the Bothnian Sea drove
in an intense snowfall that increased the snow depth in the
town of Gävle by 130 centimetres within a period of three
days. The snow was heavy, snow clearance couldn't keep up
with it, and for a time Gävle residents had to make their way
through streets where the hard-packed snow was level with
the tops of cars. Big snowfalls often happen in this way – cold,

dry air travels over large lakes or the open sea, absorbing moisture from the warmer water. The heated air rises and then cools, allowing the moisture to condense and form clouds. When these clouds come in over land and are pushed upwards, the moisture in them starts to be released in the form of snow. These snow-belt blizzards often form distinct lines along the direction of the wind, and can issue extremely large amounts of snow. Some of the biggest changes in snow depth in Sweden from one day to the next are the result of snow-belt blizzards like the one in Gävle.

The coastal city of Oskarshamn in Småland received 72 centimetres of snow on 3–4 January 1985 in a similar fashion. West of a low pressure system over the Barents Sea, very cold air flowed southwards with a north-easterly gale into a storm on the Baltic Sea. The surface water temperature was several degrees above zero, while the temperature at an altitude of 1,300 metres was -17° Celsius. The sharp temperature differences caused violent vertical air movements. Over the coast of Småland, clouds were forced further upwards. The effect was extremely localized: Oskarshamn had snow that was 110 centimetres deep; Misterhult, just 28 kilometres to the north of there, had 130 centimetres. A few kilometres south of Oskarshamn, the ground was practically bare.

The greatest snow depth at a research station outside the Swedish mountain region ever recorded by SMHI was 190 centimetres, measured at Degersjö, in the northern region of Ångermanland, on 2 January 1967. The station is only a few miles from the coast at Örnsköldsvik and Kvarken, the narrow region in the north-east between Bothnian Bay and the Bothnian Sea (literally 'neck of the sea'), which meant that for a couple of weeks a series of snow-belt blizzards thundered in over the land from the north-east. A weather observer by the name of Göta Vedin, in the nearby village of Ullånger, wrote in her diary on 1 January 1967: 'There has never been such catastrophic weather since the introduction of electric power. Wires are down everywhere because of the sleet and the wet snow. The telephone lines have also suffered the same fate. This chaos has been going on since December 17th, and is still going on as of January 1st, and it seems to be getting worse.'

This type of heavy snowfall also occurs south and east of the Great Lakes in the United States, where snow depths of more than a metre can fall in a single day. Snow-belt blizzards form over large lakes in Europe such as Sweden's Lake Vänern and Lake Ladoga in Russia, as well as in Japan, Korea and in the United Kingdom, as northerly winds descend over the Pacific and Atlantic oceans. The largest snowfalls of all categories occur in Japan and along the west coast of North America, but also in the Rocky Mountains, as in one instance at Silver Lake, Colorado, which received 193 centimetres of snow on 14–15 April 1951. A total of 480 centimetres fell on 13–19 February 1959 at Mount Shasta Ski Bowl in California. The largest total amount of new snow in a year (accumulated new snowfall) was recorded at Mount Baker Lodge in Washington State in the winter of 1998–99, totalling 29 metres. The greatest snow depth in the world (accumulated snow on the ground) was recorded at Mount Ibuki in Japan as 11.82 metres, on 14 February 1927, after a fresh snowfall of 230 centimetres – the greatest-ever known amount of snowfall in a single day.

But despite its occasional abundance, snow, with its variety of manifestations, is something many people in the world have never even been near. They've never walked with snow under their feet, never held snow in their hands, let alone stood on a pair of skis, built a snow bivouac, ridden a kick-sled – or made a snow angel.

The words used to describe snow have a certain permanence, even though the knowledge we currently have about snow is very different from what was known before. The Sámi words for snow, the words that ethnologist Yngve Ryd transcribed from Johan and Ibb-Anna Rassa in Jokkmokk, are now many generations old. Most have been around for hundreds of years. Some words probably arose with the emergence of wild reindeer husbandry, in the fifteenth and sixteenth centuries. Reindeer and their migration routes are important parts of this vocabulary. *Tsievve* is the word for hard snow that reindeer can't get any purchase on with their hooves. *Tjievttjemuohta* is snow that goes up to the reindeer's hock, while the snow that reaches up to the reindeer's belly is called *tjoajvevuolmuohta*. There could have been other words from the long period during which the Sámi lived in the forests with the reindeer, during the first centuries of the previous millennium. Like other languages, the Sámi tongue is largely lost in the obscurity of time and a culture that wasn't recorded.

Swedish words for snow also have a history. *Skare*, meaning 'snow crust', has been documented since the seventeenth century, while *klabbföre*, meaning 'snow that sticks to the bottom of the skis', came into use when skiing became common, around 1900. Words like 'crystals' and 'diamonds' appear in the very oldest texts on snow, from antiquity and the early modern period, by which point they are already established metaphors for it. This is the case in other languages, too. It's as though even early vocabulary around snow was associated with a higher sphere, in contact with a near-supernatural reality. I don't think this is a coincidence. The snow comes out of the sky, as does the rain, and they are, of course, related – but still different. Snow isn't rain. The quest of a dignified, drifting snowflake is light years away from the common, pattering raindrop, suspended vertically over its focal point on the ground. Snow is transformed somewhere on high, by a force that humans can't control – the cold. People could melt the snow into water, but the reverse – making snow from water: that they could not do. Humans own the heat, not the cold, as the French natural historian Georges-Louis Leclerc, Comte de Buffon, put it in his imaginative and encyclopaedic 1778 overview of the history of the earth and the human impacts on it, *Les Époques de la nature* ('The Epochs of Nature'). It was Buffon who discovered that history is within everything. The only thing I've added is . . . snow.

Snow's origin is magical. Even more magical than that of ice. Freezing was indeed impressive and that, too, couldn't be achieved by humans, but how it was formed was visible to the naked eye; you could watch it happen, and even supplement it by providing additional water.

But the snowflakes – who made them? And according to what plan?

Snow Anxiety

＊

SOON IT'S THE 1960S. With those long, cold, snowy winters. I remember that as soon as I got out of bed in the
mornings, I used to run to the thermometer by the kitchen
window. Then I'd rush in to wake up my dad and tell him
how cold it was. Around -37° Celsius was the norm. He
always affirmed what I said, 'Oh, still just as cold!' The cold
was our joint project. The statistics from SMHI show that I
remember it pretty well. There were weekly temperatures of
between -30°C and -40°C. Some days even lower.

One day it's -45°C. We go to school as on any other day.
The children from the villages far away get the day off,
because using transport to get to school is too dangerous.
But the kids from Söråsele, two kilometres away, come in on
their kick-sleds as usual. With white hoarfrost on scarves
wrapped around their mouths and noses, as if they were
wounded soldiers from an unknown winter war.

Sliding in the hallways in our woollen socks is sport for
us middle schoolers on these quiet, crystalline days. Or on
the reindeer-skin shoes, for anyone who had them. Then
you could give your friends an electric shock from the static.
Especially the girls, who there was otherwise no acceptable
reason to touch (except at the school dances in the darkness

of the gymnasium to the sound of The Hollies and The Shanes).

School sports take place in the snow. I tumble in big piles of snow that the snowploughs have heaped up. In the school-yard, we make tunnels to crawl in, to take refuge in during our snowball fights. In there, inside the snow, where no one can stand up straight, something feels off to me. I don't know what it is. Only that it freezes my very core.

In August of 2019, an unusual memorial ceremony was held in western Iceland. People gathered at Borgarfjörður to bid farewell to the Okjökull Glacier on the slopes of the extinct Ok volcano. The glacier site is now called Ok, too, ever since the *jökull* – which means 'glacier' – part of its name was removed because the glacier itself had become extinct. Pronounced dead by Iceland's expert glaciologists. Mourners gathered to remember the glacier, which had been there for as long as humans had been in Iceland, since the early 800s CE, and for thousands of years before that. Layer upon layer of snow, year after year. Now it was . . . well, maybe not dead, because there was nothing that had been alive and no body. But it was gone. Its existence had ceased.

Ok won't be an isolated case. Dozens of glaciers have already disappeared and, at current emissions levels, the majority of Iceland's four hundred glaciers will have melted away by the end of the twenty-first century. Already threatened is the glacier at Snæfellsjökull, the volcano featured in Jules Verne's *Journey to the Centre of the Earth* from 1864, in which the character of Hamburg professor Otto Lidenbrock descended with his guide Hans and nephew Axel in search of a lost underground world where tumultuous seas, petrified trees and wondrous ancient animals awaited, miles beneath the earth's surface.

Jules Verne's story is reminiscent of Plato's Atlantis, the mythical island that had sunk into the sea, but also of the new dynamic concept of the earth that became increasingly well-established by the middle of the nineteenth century. Darwin's theory of natural selection had been published just five years earlier. Theories about an ice age that had paralysed the northern hemisphere, wiping out the mammoths and other ancient species, had been around for more than twenty years. Perhaps most importantly, at almost exactly the same time that Jules Verne was working on his novel, geologists – practitioners of a relatively new profession – were beginning to consider whether they needed to introduce another epoch into the earth's timeline. That is, a new top layer in the stratigraphy (from *stratos*, meaning layer).

The proposal was made by the French geologist and palaeontologist Paul Gervais in 1850. Gervais was mapping mammalian fossils when he was struck by the idea that a case might be made for a specific term for the geological period that included the presence of human beings. He published an article on the subject in a journal issued by the Montpellier Academy of Sciences and Letters. His request wasn't immediately granted; it took more than a century, as it turned out, but in the 1980s the new post-glacial period was finally recognized – under the name proposed by Gervais, the Holocene. By that time, geologists and palaeontologists knew that there had been at least a dozen ice ages, which had come and gone over millions of years. The greater depth of ice ages behind us doesn't make it any less important to emphasize that the post-glacial period, the one in which modern humans are living, is nevertheless a different time.

Nowadays, each newly discovered geological period has a stratigraphic marker, a round brass plaque inserted into the earth's crust at a point where the characteristics of that period become clearly visible. For the Holocene, there's no such 'golden spike' in the visible environment, however. That's because stratigraphic experts concluded in 2008 that what is most representative of the end of the last ice age is the layer in the Greenland Ice Sheet that corresponds to snow that fell approximately 11,700 years before the year 2000, which is the year when the shift to a warmer climate presents itself most clearly. That layer, encapsulated in a sample 'drill core' extracted from the ice sheet in the year 2000, is found at a depth of 1,492.45 metres. That thousands-of-years-old snow has been preserved there, below 1,500 metres of snow accumulated at a rate of around ten centimetres a year, since it first landed on the glacier almost 12,000 years ago. This critical ice sheet contains the boundary between the Holocene and 'civilization' itself, and is stored, somewhat unglamorously, in a cardboard box among other segments in a large freezer archive filled with sample cores at the Niels Bohr Institute at the University of Copenhagen.

It's worth pausing a bit to consider history. After 2.5 billion years, snow is making its return as a means of dating Earth's past. The first time was 'Snowball Earth', when snow first covered the planet. In our own time, for those of us living in the present day, after this long loop of billions of years and more than a hundred geological eons, epochs, periods and ages (geologists work with a lot of different concepts regarding the eras of the earth), snow is back in the stratigraphy – at the very moment that it's starting to disappear. The end of the last ice age is nothing but the beginning of a warmer period. The Holocene has been climatically fairly stable for almost twelve thousand years. But over the last century, the Holocene has shifted into the 'Anthropocene', the age of humans. And in the Anthropocene, we're the ones who are warming the planet, and what we have seen so far is just the beginning of a rise that is certain to continue for a long time, probably pushing the temperature up by 2.5 or 3 degrees, maybe even more. That's what's causing the snow to leave us.

That's also what necessitates the mourning ceremonies at glaciers. At the tombstone during the solemn ceremony for Ok, Icelandic prime minister Katrín Jakobsdóttir and former UN Human Rights Commissioner Mary Robinson were joined by about a hundred other people who wanted to show their commitment to preserving cold in the world and, accordingly, Iceland's and the world's glaciers and snow – and all those who depend on these elements, in some cases for their survival. There was a short text written on the tombstone: 'A letter to the future', by the author Andri Snær Magnason. The text ends with a message to future generations:

> This memorial acknowledges that we know what is happening and what needs to be done. Only you know if we did it.

The plaque also bears the inscription '415ppm CO_2'. This was the average level of carbon dioxide in the atmosphere at the time of the ceremony. The figure is the result of a measurement carried out at a research station at the Mauna Loa volcano in Hawaii. At the same time – without saying so explicitly – it is a reading of time itself. The annual – and daily – increase in greenhouse gases is a literal chronology that translates into a corresponding impact on living and dynamic materials all over the world. One could also say it's a countdown – for the species that are becoming extinct and for our civilization, in which we have become accustomed to progress but are increasingly needing to measure our time in losses. The French geologist who proposed the new post-glacial era couldn't have dreamed of anything like this in even his wildest imagination. Nor could he have imagined that only a few decades after the introduction of the Holocene, another new epoch, the Anthropocene, would be discussed to signify mankind's domination of the elements. Including snow.

Since the death of Okjökull, several similar memorial ceremonies have been held. A month later, in September 2019, a few hundred mourners, dressed in black, gathered at an altitude of 2,700 metres at what had been Switzerland's Pizol Glacier, in remembrance and to bid farewell. Only an area the size of a couple of football pitches remained of the Swiss glacier near the border with Austria and Liechtenstein. 'We're here to say goodbye to Pizol,' said glaciologist Matthias Huss, to the sound of long alphorns. 'We estimate that five hundred glaciers have vanished from Switzerland since the 1850s,' he continued, adding that the deglaciation has been dramatically fast. 'Pizol has lost 80–90 per cent of its surface since 2006.'

In August 2020, it was time to mark the passing of Clark Glacier in Oregon. This took place outside the State Capitol building in Salem, a monumental Art Deco-style edifice with a white marble facade. A dark coffin containing some of the glacier's last meltwater was brought there and stood in stark contrast to the pale stone. The aesthetic was inspired by a photograph of Ruth Bader Ginsburg, the human rights lawyer and United States Supreme Court justice, who at the time had just been lying in state at the US Supreme Court in Washington, DC. Ginsburg had often adorned her black judges' robes with a white lace collar. Mourning clothes, but also the image of a mountain with its glacier. Ginsburg, a national heroine known for her judgement and compassion, left her mark on the mourning ceremony for a glacier in Oregon, on the other side of the great continent.

The glacier was named after explorer William Clark, who, along with Meriwether Lewis, crossed the continent to the Pacific Northwest at the directive of President Thomas Jefferson, as part of an expedition lasting more than two years to map the central and northern parts of the present-day United States. These lands had become the property of the United States through the Louisiana Purchase of 1803, when it bought these vast territories from France. Much later, Lewis and Clark each gave their name to a glacier on the Cascade mountain range in Oregon. On maps dating from the 1900s and well into the 2000s, you can see the two white surfaces with their underlying blue elevation contour lines. Satellite images from 2020 show the new reality. Of the Clark Glacier, which has greater exposure to the sunnier south-west, only isolated fragments remain. Even the once-continuous Lewis Glacier, despite having a more advantageous position in the north-east, is now shrunken and in separate pieces. Doomed to die. Everything frozen melts.

*

In May 2024, it was time to mourn once again. Only two hectares of Venezuela's last glacier remained, now completely lacking its accumulation zone, the area in the upper part of a glacier where it's cold enough for new snow to pack and 'feed' the glacier with fresh ice. Over the previous three years, the glacier had lost half its surface area and been downgraded to an 'ice field'. Not long ago, Venezuela had half a dozen glaciers, but it will now be able to claim a notable and novel honour in world history: that of being the first country in contemporary history to lose its last glacier. That is to say, to have lost them all to anthropogenic warming.

The final Venezuelan glacier was named after the German explorer Alexander von Humboldt, who in the early nineteenth century had been able to evaluate the impact that European colonization had already had on the South American continent, through extensive scientific measurements. Little did he realize that his own name would literally disappear from the surface of the continent due to the impact of human beings. When it was larger, the Humboldt Glacier wrapped itself around Venezuela's second-highest mountain, Pico Humboldt, nearly five thousand metres above sea level. Its former name had been *La Corona* – the Crown. It crowned this tropical country with snow. The ultimate irony, of course, is that Venezuela is an oil-producing country whose export earnings come at this high price. But Norway, Russia, China, Australia and other fossil-fuel-producing countries also contribute, as do virtually all emitters. The rich contribute the most, but the number of us who in some way contributed to this is enormous. Our modern lives have turned us into snow melters, whether that's who we want to be or not – 'For the good that I will to do, I do not do; but the evil I will not to do, that I practice', to quote St Paul. Next in line to form a tragic group of countries that have lost their last glaciers are Indonesia, Mexico and Slovenia. British glaciologist Caroline Clason sees the loss of the Humboldt Glacier as a 'reminder of why we [glaciologists] do our job and what's at risk in these environments and for society'.

Our preoccupation with snow disappearing is a new phenomenon, a feature of our emotional state at a moment that hasn't existed before – at least, not for the snow in and of itself. The fact that we have an overall picture of the quantity and breadth of snow throughout time is something previously unknown to humanity. It's only now that we can have this kind of understanding, arising over the last few decades; that we can state that snow likely has never before vanished in the northern hemisphere to the extent it has done since around 2000.

There have been warm periods in the past. Northern Europe's mild winters in the 1930s, with warm, moist winds all the way up into the Nordic countries, were the subject of mournful commentary. Glaciers melted, migratory birds ceased their travels and slush in the Nordic capitals periodically reached depressing levels. There was concern for the small jewel that is the Helags Glacier in Härjedalen in Sweden. What would happen to it? From Norway there came 'alarming information' that 'the high mountains are losing their most beautiful summer clothes, the white snow and the blue glaciers'. But the overarching perspective was the complete opposite. If climate change, which at the time was portrayed as something entirely natural, was now actually the cause of the decrease in mountain snow – which was already then being disputed – surely that wouldn't be anything to shed tears over?

Even if it were to get warmer in the Nordic countries, that wouldn't be a disadvantage – quite the opposite. Agriculture would produce higher yields with increased heat. Summers would get warmer – and who could complain about that? Nordic folk and Britons would get a little sun on their anaemic, pigment-deprived bodies. One of the leading climate researchers of the time, Stockholm geographer Hans Ahlmann, issued a summary of what he called 'the current climate improvement' in an article in the *Svenska Dagbladet* newspaper in 1938:

Plants bloom and fruit in more northerly and higher altitudes than before. Such that in Norway, barley and even wheat, have ripened at much higher altitudes in recent years than ever before in living memory. It can be said that the upper limit of grain cultivation in Norway has risen to a few hundred metres. Certain species of fish have [. . .] extended their migration further north than before, and migratory birds have cancelled their journeys from Arctic areas to more southerly regions, preferring to remain throughout the polar night on those coasts from which the severe cold and frozen waters previously drove them away as early as September.

These were tangible and, at the time, mysterious changes, but few people complained about them. Yes, if there was too little snow on the ground, forestry – which used horses and sleighs – could be disrupted. The spring rivers might also be affected, electricity production reduced and the water-flow too low for critical timber transport. But these were marginal voices, raised in industry journals that few people read.

The concerned voices didn't make it into the public debate. But those voices singing the gospel of climate improvement, which recurred in the Scandinavian mass media for a couple of decades until around the mid-1950s, did.

Not even the reanimated theory involving human-made climate change through greenhouse-gas emissions could dampen the enthusiasm. 'It's the carbon dioxide doin' it!' said the headline in the daily newspaper *Aftonbladet* in 1954, as the CO_2 hypothesis re-entered the debate, the idea having been introduced in 1896 by Stockholm chemist Svante Arrhenius. Because Swedish and Norwegian research within these areas was influential, the message reached an international audience. Stockholm-based Professor Ahlmann conducted his research throughout the North Atlantic region – in Lapland, in Jotunheimen in Norway, on Svalbard in the Arctic Ocean, on Vatnajökull in Iceland, in Greenland and, after the Second World War, in Antarctica, with the help of younger colleagues. He made lecture tours in Norway, the UK and North America, making headlines about the new warming era. It was basically considered a good thing, at least according to most experts. Even Ahlmann himself opined, 'What's been negative for the glaciers, somewhat catastrophic, is something advantageous and beneficial for mankind and other living creatures.'

On a more personal level, though, he mourned his glaciers. He had studied them for such a long time and become so close to them that he had started having feelings for them. It was Ahlmann who voiced the earlier quote, that 'the high mountains are losing their most beautiful colours, the white snow and the blue glaciers'. He was not alone in his vacillation. Yet there were locales and communities, as well as individuals, who were also very fond of snow, but had a vague sense of unease. An early, nebulous but still palpable nervousness that the rapid disappearance of snow in the 1930s might not be solely positive. That it could be the start of something very worrying.

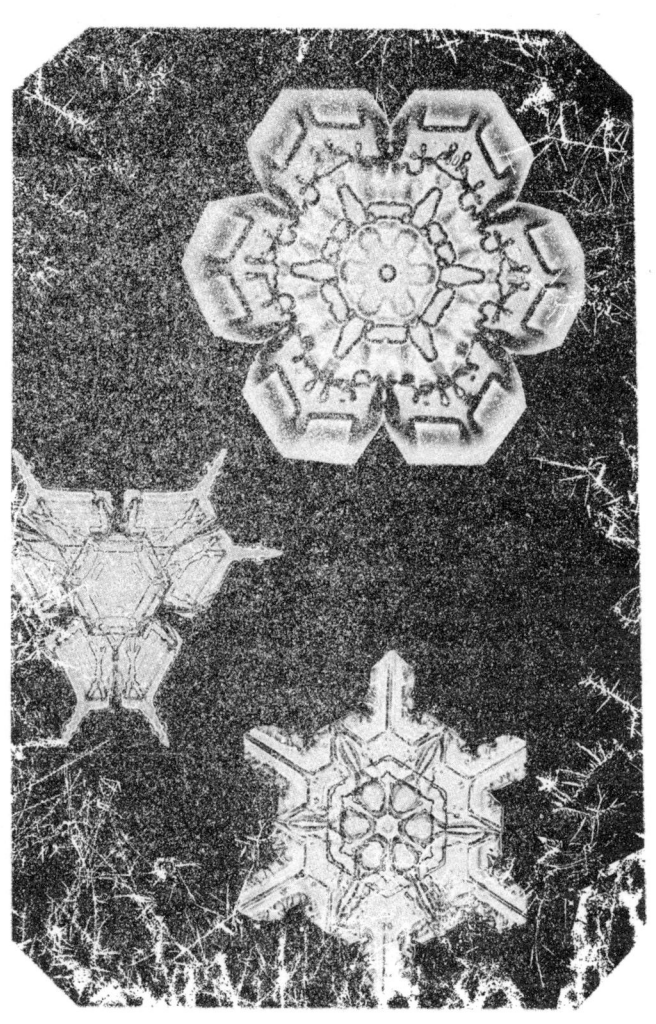

Amor nivis

*

APRIL IN THE RONDANE MOUNTAINS IN NORWAY. An early warm spell has melted the snow on the mountain plateau. Meltwater pools in black puddles between the tussocks near our camping site. We are three young adults about to ski through the Norwegian massif, at some point in time well into the 1980s.

The sun disappears behind an esker. Soon the cold tightens our cheeks. The puddles around us start to freeze. The smallest one first. A delicate trim of ice on the edge of a leaf marks the point of crystallization. Then, in just the blink of an eye, comes a wave of expanding crystal that, with a small snapping sound, covers almost the entire puddle. First one puddle, then another and another. Then dozens of puddles freezing in the small valley, each giving off its own cracking sound as it does so. Crystal shards play a calm taps there on our plateau. The larger pools, the ones that are almost ponds, freeze last of all. The deeper water delays freezing from below. When the inevitable happens, it's with such powerful force that it makes crises easier to understand.

This twilight in Rondane lodged itself deep within me on that trip, taking up permanent residence. I've thought of it again and again throughout my life. I'd never experienced ice forming in front of me like that before. What was it that I was feeling? Maybe there's something about seeing the forces of nature, the whole solar system, at play in a condensed time-scale that brings everything into focus. Something like the astronomer Johan Carl Wilcke must have experienced when making his ice hexagons in eighteenth-century Sweden, with soap and water at his open window in winter. But I didn't have that image available to me during that moment. I didn't know anything about crystals.

Perhaps it's in that moment that I realize there are forces which can destroy everything. Or create everything. In some odd way, at this second I'm standing in the centre of all Creation. Where everything intersects. The sun setting, shadows falling, the cold increasing, ice forming. It all happens in an instant.

Maybe it's because it's an image of the nature of existence. The state of existing. An existence that's always changing.

This is the way that crises begin, right in the middle of everyday life. Seemingly mysteriously. Small signs; a trail of shadows, a chill in the air. Soon, one's entire being is shaken by a mighty storm. Whole communities are uprooted. A crisis can be postponed, but rarely prevented. Under certain conditions, ice forms. This cannot be changed.

The cause of a crisis lies in the past. The same goes for revolutions: no one believes they're possible until they're in progress and you realize that it couldn't have gone any other way. The transition can be a crystal. Maybe a snow crystal. These incalculable forces . . .

And now we are ourselves just such a force. No longer is it the sun, shadow and the cold that come to us through providence and chance and the immutable laws that have been built into Creation since its first moment. I think of this long-ago moment in Rondane as a harbinger. A way of making me receptive to what our new conditions are. And those conditions are *us*.

The conservative philosopher Edmund Burke felt horror and disgust at the French Revolution, at the same time calling it 'the most astonishing event that has hitherto happened in the world'. Burke's fear lay in his realization that society contains forces that are capable of anything. A social order that has taken generations, centuries, to form can be overthrown in an instant. The question, then, is not whether societies contain forces that can transform the world. The question is what these forces will be used *for*.

I now consider this sunset in Rondane as something that allowed me to understand what can happen when the time is right. It can end in any of a number of ways.

The world we're creating can turn us into the kinds of people Dante wrote about – the ones who've committed the mortal sin of gluttony and reside in the third circle of hell; those who stand in stinking mud, whipped by eternal ice-cold rain. 'Gross hailstones, water gray with filth, and snow / come streaking down across the shadowed air; / the earth, as it receives that shower, stinks.' We are the ones who've created an overheated Earth and unleashed forces more powerful than a thousand suns.

This is what's usually termed an epiphany. A visit from an angel.

The Angel of History. The Angel of Snow.

*

Amor nivis – Latin: the love of snow. I think I can say by now that this kind of love exists. But it hasn't always existed. It must be given a form and meaning. You could say that it's something that has to become a concept. It's making its way, trying to carve out a place. Not everyone is touched by this love. In Sweden, ultimately it is bound up with our feelings about our common home under the roof of our snow globe.

This kind of love tentatively emerges in the late eighteenth century as part of the spiritual and intellectual current that a Swedish literary scholar once called the 'Romanticism of the Enlightenment', a little over a century ago. This was followed by a wave of sentimentalized patriotism involving snow, and in the nineteenth century the worship of snow grew further and became more intimate. Adopted by the family culture of the growing bourgeoisie, it can be seen in the newspapers and popular magazines that were read around the kerosene lamp in Sweden and the Nordic countries, but also in Puritan New England on the other side of the Atlantic. Perhaps it's a Lutheran disposition, grounded within the Reformed Christian countries of northern Europe and their colonies – areas abundant with snow.

Snow worship then turns into snow propaganda, aimed at children in particular, with songs, rituals and winter games that are considered an essential part of childhood. 'Oh, Look, It's Snowing!' is a cheerful song by Swedish composer and church musician Felix Körling that conveys a sense of welcome surprise. He wrote it in 1913, but even at that time a snowy winter wasn't a given. At least not for Felix Körling, who came from the Kalmar region and worked in Ystad and Halmstad – all in the southern part of the country. In the last stanza of the song he exclaims, 'Hurrah for winter, hurrah for winter, it's finally here!' set to the music of Zacharias Topelius's 'Schoolboy's Winter Song': 'Oh, now it's winter and my skis are sliding'. I remember it myself from school.

With our traditional *kakelugn* (tiled stove) glowing in the corner of the schoolroom, we learned that snow was good. Part of that goodness involved faithfulness. That, too, had something to do with the snow. Just like the song my mum sometimes sang at home about the fir-tree in the forest in winter: '*O Tannenbaum, O Tannenbaum, wie treu sind deine Blätter / Du grünst nicht nur zur Sommerzeit, nein auch im Winter / Wenn es schneit.*' ('O Christmas tree, O Christmas tree, How lovely are your branches! Not only green in summer's heat, But also winter's snow and sleet.') When I was little, I didn't understand the words. But later I recognized that it was about mortality. It was about seasons coming and going with secure regularity. A message that was appropriate for children: that it would snow every year. And that under the snow stood the fir-tree with its faithful green branches, on which the whiteness would fall. The fir was a warm tree, under which a hare could sleep. The snow itself was also warm. This, too, was important. A nation – especially a nation located by the Arctic Circle – needs warmth if it is to be a community. Indeed, an illustration of a fir-tree, by the artist Olle Hjortzberg, featured on the cover of the primary school reader *Läsebok för folkskolan.* Under the tree was a little cottage, a home, with heart-shapes in its shutters. There's an adage on the cover of that book that reads: 'Listen to the whispering of the fir tree at whose root your nest is built.' The tenth edition, the one that I own, came out in 1919 and still carried that same image. The fir tree was shelter. Covered with snow, it was also silence, stillness. Trees and snow worked together to generate a sense of community. Trees and snow have always existed. But this new national combination of the green and the white was a discovery, a visual innovation.

This visual pairing was reinforced by science in some intriguing ways. At the same time that artists began to take an interest in snow as a component of national identity, a physics thesis was published at Uppsala University in 1904, entitled '*Om värmeledningsförmågan hos snö*' – 'On the Thermal Conductivity of Snow'. In the same year that official measurements of snow depth began in Sweden, this study lays out why snow is such a good insulator – as good as sawdust, in fact.

Soon, snow is associated with a specifically Nordic version of healthfulness, conveyed by style icons of the time such as Carl and Karin Larsson. The artist couple from the central Sweden region of Dalarna didn't only promote idyllic images of summer. They presented an equally idyllic winter, replete with images of mittens, skis and kick-sleds. Swedes become a winter people. Like a snow snake, winter patriotism winds its way into the verdant summer meadow. It wraps itself around the iconic midsummer pole with a confident, exhilarating vigour, literally replacing it during the quieter, whiter season. A wider trail opens between Sälen and Mora, ritualizing since 1922 the mythical history of snow through the cross-country Vasaloppet ski race, which was inspired by the flight of the sixteenth-century future king of Sweden from his Danish pursuers. The race has subsequently become a Swedish passion and reinforces a wintry strand of the Swedish national self-image.

By now, the theme exists in several countries. Norway takes it the furthest. Snow expert and polymath Fridtjof Nansen does everything that there is to be done with snow. He becomes a pioneer in the genre of ski literature, reporting from the Hardanger mountain plateau. He undertakes trips, sometimes lasting several days, with his wife Eva, a member of the powerful Sars family. Fridtjof and Eva meet, fittingly enough, one winter day at the ski resort of Holmenkollen. She'd had a fall. He sees her two dark eyes under her frosted hair and lifts her out of the snow. They start skiing together. After Nansen's famous Greenland voyage and newfound hero status, they get married.

Eva Nansen loves both snow and skiing. But the knickerbocker-wearing male elites don't think that women and snow should mix. In 1893 Eva Nansen writes an unconventional newspaper article: a defence of women's skiing as a route to health, beauty and independence. How could men be against that? Fridtjof Nansen conducts research on snow and ice, and constructs theories on *friluftsliv* – championing outdoor life as the most ideal kind, and as the Nordic contribution to civilization. Wearing a knitted sweater, he poses, *contrapposto*, on his skis as naturally as the leaders of other countries do on horseback, and in so doing cements his persona in both image and word, becoming a commanding leader of the movement that would ultimately lead to the dissolution in 1905 of Norway's union with Sweden. This union had begun with Denmark ceding Norway to Sweden under the Treaty of Kiel in 1814 – against the will of most Norwegians.

I carry Rondane inside me like an unanswered question; a fear with no tangible basis. Even a limited nuclear war could drastically lower the earth's temperature by enveloping the planet in a permanent cloud of dust. In 1983 the American astronomer and media figure Carl Sagan coined the term 'nuclear winter'. There was talk at the time of 'twilight at noon'; the phrase was reminiscent of Arthur Koestler's 1940 novel about the Stalin era, *Darkness at Noon*. But this wasn't about totalitarian terror, it was about how nuclear war would create a planet in perpetual twilight; a shadow existence, for those who happen to survive the war itself. There had been an earlier article published in the journal *Ambio*, in 1982, entitled 'The atmosphere after a nuclear war: twilight at noon', written by the meteorologist Paul Crutzen and John Birks, an expert in atmospheric chemistry. Crutzen, a Dutch native who'd done his training and research at Stockholm University, would win the 1995 Nobel Prize in Chemistry for his research on the ozone hole.

I remember the subject of the nuclear winter. This was a different kind of winter, distant from those I'd experienced as a child. I couldn't take it in. Nonetheless, something about it shook me, wordlessly, to the very centre of my being.

That's when I discover the snow. It's September 1984, just a few months after Carl Sagan has testified to Congress about the nuclear winter. I'm in Toronto – it's my first trip across the Atlantic. I'll be here for a month. It's a turbulent time in my life. I am emotionally dried out and the anxiety fits I've been suffering are getting increasingly worrying.

Museums and art always make me feel at home amid everything that's unfamiliar, so I visit the Art Gallery of Ontario. Earlier that year, it had put on the exhibition *The Mystic North*, including Nordic painting from the turn of the twentieth century, and I get hold of the exhibition catalogue. Harald Sohlberg's painting *Winter Night in Rondane* is among the exhibits, and it casts a spell on me. Two massive snow-capped mountains – in the middle of a new, soon-to-be independent Norway – arch fertilely and brazenly against the deep blue of the night sky. Between the massifs, the diamond that is Polaris, the North Star, a sparkling piece of jewellery on the neckline of the young nation. In the foreground, tree branches that give the painting depth and a sense of proportion against the mighty mountains.

Sohlberg did several versions of the painting during the first decade of Norwegian independence. The one I saw in Toronto is the earliest of the paintings, from 1901. Sohlberg visited Rondane for the first time during Easter 1899. He was a sportsman who skated, sailed, played football and, of course, skied. Having been invited on a trip to Rondane, he accepted. He couldn't have imagined what would happen.

On the train home from his trip, Sohlberg is already drawing the mountains on the back of an envelope. In 1900 he returns to Rondane, settling in for a long stay in the region. In February he makes a charcoal drawing of the two mountains, moving his vantage point to get the perspective he wants. He now wishes to see the mountains from a higher position, so he again changes his viewpoint, continuing to work on this first painting of the mountains, while also starting on a larger version.

But in 1901 he pauses the work. Art historian Øyvind Storm Bjerke, who studied Sohlberg's process for the creation of the painting, attributes this interruption to the artist's marriage to Lilli Hennum. In the winter of 1902, the couple move to Røros, where their son Harald Jr is born. Sohlberg intends to complete the large painting, which he has brought with him to their new home, but first works on his famous winter paintings of the mining town, which are rich in detail and almost photographically accurate. It is not until 1911 that he returns to Rondane, and again in 1913. Only once he is in the mountains, his eyes directly on his subject, is Sohlberg able to do anything really substantial with his Rondane painting. He continues to work on the trees in the foreground, black and sprawling against a dark winter plain, which are reminiscent of trees painted by German Romantic Caspar David Friedrich almost a century earlier. The colours are incredible, spanning the entire palette, especially the hues between green and violet. The mountains grow, rising up out of the ground like a geological birthing taking place in the snow.

From accounts of his comments about them, we know that Sohlberg was pleased with his mountain paintings and that he considered them important. He was quite preoccupied with them, in fact, watching over them for the next fifteen years. Maybe he was possessed, with the kind of obsession explorers experience. Was my own preoccupation with the painting so intense because I suspected that Sohlberg had been as moved and taken by these mountains of snow in the winter night as I had been by the crystallization I'd witnessed in a place not too far from there? When he resumed the project, in 1911, continuing to paint the same subjects, he found himself at a crossroads between a number of forces. What he saw before him in such intensity was no longer just the mountain, but a country that had just been born, gaining its peaceful independence from Sweden in 1905, and was now taking physical shape right in front of him. It stretched from the Russian border on the Arctic Ocean all the way to the fjords in the south-west. Standing before this mountain range, he saw dimensions of it and of Norwegian nature that few before him had done so as profoundly.

Sohlberg's painting is typical of the movement towards snow during the breakout period of Norway's winter nationalism. Artists are leading the way. There is snow-focused poetry, and many snow-oriented novels, as in the other Nordic countries, too. The prevalence of the word 'snow' grows through sports and the greater embrace of *friluftsliv*. Nansen's books *On Skis Over the Mountains* (1879) and later *The First Crossing of Greenland* (1890), which included the experienced Sámi skiers Ole Nilsen Ravna and Samuel Balto, from the town of Karasjok, are part of the long, rhetorical preparation the people of Norway undertook for their long-awaited break from Sweden. It was on skis that Norway had militarily defeated the Swedes at the Battle of Trangen in Hedmark in April 1808. This was followed by the unhappy 'indoor era' of the union with Sweden: 'the country's healthiest and best young people have become lax, they are deprived of effort', wrote Colonel Henrik August Angell. Another soldier, Oscar Wergeland (the younger brother of national poet Henrik Wergeland) published *The History and Military Use of Skiing* (*Skiløbningen, dens historie og krigsanvendelse*) in 1865, in which he complained that interaction with snow had completely vanished. Legislation was needed to ensure that Norwegian farms, including those 'in the actual snow villages', were equipped with skis.

In Sweden, things were hardly better, although there were fewer complaints. Sports enthusiast Viktor Balck, born in 1844, notes in his 1929 *Memories* that 'skiing was completely unheard of' in his youth, and literature from the snow-deprived province of Blekinge generally confirms this picture. Carl Hasselberg writes in the 1860s that in the Jämtland region any classmate who could ski off-road would be considered unique. Against this background two national ski associations, the Norwegian one from 1883 and the Swedish one from 1892, emerged for the 'promotion' or 'advancement' of skiing. Skiing with, by and for the people had hardly existed before that. It had to be introduced and, where it existed, enhanced. Even the 'snow countries' didn't identify in that way or see themselves as such until fairly recent times.

Then it all changes rather quickly. Norway's new sense of nationalism is built out of snow. Snow depth becomes something of interest at the same time that the image forms of a snowy national landscape. Norway is ahead of Sweden in terms of snow-measuring stations, starting in the 1890s, a full decade before the Swedes. If the measurements provide breadth and depth, art creates the national iconography of winter. Using brushes of the finest squirrel hair, painters caress forth images that are to become the nation's altarpieces. This happens in other parts of the snow-filled world as well, but it's more prevalent in the Nordic countries than in most other places, maybe because snow was so common throughout the region. The exception was Denmark, but it had control of what are now Iceland and Greenland, and therefore still had a right of residence in the world of snow.

Snow hadn't been that prominent in earlier studio and portrait paintings. Now it comes to the fore: expressive, filled with nuance, a signature of Scandinavian art in the decades either side of the year 1900. Gustaf Fjæstad's paintings of winter trees in the Swedish central region of Värmland are abundant with snow, including soft blankets of white extending down to touch the dark waters of a river. Snow for Fjæstad isn't white and smooth, but spotted with blue, with pink; sometimes turning into pure pointillism. Balls of hoar frost on lake ice are a recurring speciality of his. Fjæstad himself moved out into the snow, founding an artists' colony at Lake Racken near Arvika, not far from the Norwegian border. He was influenced by Bruno Liljefors and Carl Larsson, both of whom he worked for as an assistant in the 1890s, and was also inspired by Japanese paintings of nature. Fjæstad emerges as a multitalented 'winter Swede', typical of the era, competing in ice-skating at a high level and, in 1891, setting a world speed record for an 'English mile', at competitions in Hamar, Norway. He's also a cyclist. Athletic. Practises outdoor activities well into his later years.

As with Sohlberg, Fjæstad's most powerful paintings of snow are produced after the turn of the century. More than any other artist, he lets the snow blanket everything underneath in a fairy-tale world of shapes, just as he has in the painting on the cover of the book you are currently holding in your hand. His embrace of snow, the deep, blanketing display of it, the subtle high-pitched ring of snow in the National Romantic movement, peters out around the First World War. Then Helmer Osslund from the Swedish region of Ångermanland takes up the mantle, moving northwards into Lapland. By then, the peak in the national painting trend of winter-worship has passed. Snow has gone from the foreground, but not completely. It remains in the white flakes of Osslund's brown, autumnal mountainsides or amid the mighty mountains on the coast of the Bothnian Sea. The snowfields are part of a dizzying panorama of colour, testament to the beauty and richness of the Norrland region, which Osslund both came from and celebrated. In Lofoten in northern Norway, Anna Boberg is painting fjords and fishing villages crowned by the dense snow of the high mountains, which reflect the low Arctic light in an air that seems to tremble with microscopic diamond dust – the critic Johnny Roosval calls Boberg's snow 'candied'. We get art from snow – painting with snow as its theme – which still hasn't been studied enough, given its significance.

The master is Pekka Halonen from Finland. His gnarled pines are covered in a thick grey-white snow, as creamy as Sicilian *gelato*, proving that barren Finland isn't barren at all. Halonen was a committed nationalist. His early career, in the 1890s, followed a typical pattern for Nordic artists, with a string of fellowships in Paris interspersed with periods of nature painting in his homeland. He was influenced by Jean Sibelius's patriotic compositions and painted scenes from Finland's national epic poem *The Kalevala*. Snow scenes grew in importance in Halonen's work, with some of the most iconic being created in the 1910s and 1920s. Halonen also depicted local traditions such as doing laundry out on the ice and lynx hunting. At first, the snow is simply the background, but soon it comes to the fore, forming motifs in an increasingly austere, pared-down style showing Japanese influences. Another Finnish painter of snow landscapes is Akseli Gallen-Kallela, famous for his paintings of the Imatra Rapids, which were a major nineteenth-century attraction and destination for members of affluent society from both Finland and St Petersburg. In January 1893 Gallen-Kallela painted on-site, producing iconic winter landscapes that further strengthened the reputation and nimbus of the place. He also designed advertisements for the Imatra Rapids' tourism company.

The popularity of snow in 1900 is something of a turning point. Nordic artists had long travelled and studied in

France and Italy. Now, for a brief moment, the transitional dawn or dusk light of snow in the Nordic countries draws more southerly European artists northward. Impressionist Claude Monet, the master of light himself, travels to Bærum, on the outskirts of Oslo, and paints the Kolsåstoppen mountain day after day, over a period of a few weeks in the winter of 1895. In total, he completes twenty-eight paintings. He gains critical knowledge that he takes home with him to Giverny, and his now-famous water-lily pond in his gardens, which he starts painting a couple of years later. Snow becomes an asset, and the Nordic region something to look up to.

The snow is everywhere, but I haven't really *seen* it. I've made use of it without understanding it. It's something that just comes to me, presents itself. First in Rondane, then in Toronto. I have to be somewhere else to properly see the world in which I've grown up. Even now, the ability to see it isn't of my own doing. It's the artists who do that work for me. I realize that what the Nordic painters had done was to *discover* what the familiar really was: something surprising and astonishing. Worth standing in front of and reconsidering. I also realize, somewhat tentatively, that my own upbringing, which once seemed so commonplace, was part of the self-same mystery that the Toronto exhibition evokes. I myself am part of that mystique. It's we in the Nordic lands who are the sagas and mythology. It's beginning to dawn on me that the snow has a voice that it uses to speak to us, with many important words that we've forgotten.

The Art Gallery of Ontario is part of that conversion, though I'm not yet clear about what the implications are.

I consider my home country to be the soul of reason, and therefore the very seat of boredom. But really, all of it is magic. Each life rests on the lives of others.

At a certain wavelength, one human being always has some understanding of another, her joys, her suffering. So what does the snow do? It makes us notice her. No one can help but see a solitary person in the snow.

Snow is important in the Nordic countries, but we're neither foremost nor alone in that. In the early twentieth century, snow's transformational mission made its way all the way to Hollywood. The new medium of film allowed snow to soothe, comfort, forgive, love. Alongside music, it carried the emotions of the romantic comedy, which have since been made as a way of getting people into cinema theatres during the month of December, when attendance tends to be poor. Snow's dramaturgical function is soon codified. The most beautiful snowfall comes when the power of chemical attraction has triumphed over all obstacles and misunderstandings, and it's clear that the lovers are going to be together. Up to that point, it's usually slushy. The film zooms out as the main characters zoom in on one another.

There's also something about snow's ability to defy gravity. To be able to fall upwards – and yet sooner or later reach the ground; the same attribute that a light spring coat can have. Or the heavy snow globe containing a tree or a *tomte* (house elf), and snowflakes on the bottom. When I turn it upside down in my childhood bedroom, it starts to snow – but upwards. I have to turn it right side up again before the snow falls the way it should. Its passion is carnival-like, turning the world upside down, making a fool of an emperor and the poor man a king. The snow is like that every day, an invitation to change and to lose oneself in emotions. In Tomas Tranströmer's 1962 poem 'C Major', snow swirls in the air. It's a symbol for happiness: 'The night shone white [. . .] Passing smiles – everyone smiled behind upturned collars'. The man in the poem walks faster, even though it's the middle of the night – 'The whole town sloped.'

Snow can frame life, as Tomas Tranströmer writes forty-two years later, in his poem 'Snow Is Falling':

> The funerals keep coming
> more and more of them
> like the traffic signs
> as we approach a city.
>
> Thousands of people gazing
> in the land of long shadows.
> A bridge builds itself
> slowly
> straight out in space.

Care of Snow

*

IN THE DARK OF AUTUMN, SNOW ARRIVES. The day when the first snow falls is the most beautiful day of the year. It descends slowly, like in the snow globe I turn upside down.

The next day is also beautiful; now the snow on the ground shimmers like a night light that God has set out to comfort people through the long hours of darkness. Later in the winter, on days when it has snowed a lot, I usually wait for the snowplough. Yellow with red plough-blades. It sounds like a moving earthquake. When it approaches, I jump onto the plough embankment, lying on my back. The rumble grows and, as the plough thunders past with a deafening roar, the accumulated mass of street snow floats down over me, burying me in soundless, heavenly weight.

A few seconds of peace. It's just me in the world, and I don't have to do anything.

Snow does a lot of work on behalf of the climate. Just like the oceans, which absorb 90 per cent of the excess heat generated by the carbon dioxide we emit. Snow, meanwhile, reflects sunlight: up to 90 per cent of the light that hits a snow-covered surface is reflected back into space. The albedo, or whiteness, of snow is 0.9 – which is very high. Soil, trees and plants have an albedo of between 0.1 and 0.3, meaning they reflect back less than a third of incoming sunlight. The oceans reflect less than a tenth.

If there were suddenly no snow at all, the area covered by snow today would absorb several times more energy than they do currently. The continents and oceans would warm up even faster than they already are. Apart from forests and rainforests, few things have a greater impact on continental temperatures than snow albedo. The biggest change in the long-term decline of snow in the northern hemisphere is to spring snow, left in April and May; this snow is disappearing the fastest. This results in more months of strong solar radiation on bare ground and an ensuing decrease in the total albedo effect of snow, which in turn drives warming via so-called 'positive feedback', the effects of which, in this case, are exclusively negative.

Snow affects food production and, in some areas, can mean the difference between famine and plenty. Snow poses risks, both of avalanches and traffic problems. It also determines the availability of water. In areas where the water supply comes from snow-covered areas, a shortage of snow, known as a snow drought, can be a major concern, as it points to a later water drought, and in turn cause shortages in energy supply where there's a reliance on hydropower.

Periods of snow drought are increasing in Europe, North America and Russia, while the extent of snow cover is decreasing in most places in the northern hemisphere, mostly as a result of climate change, though the rate varies depending on local and climatological factors. On average, we're talking about a reduction in the number of 'snow days' in Europe of between three and five days per decade over the last sixty years. The corresponding figure for North America and Eastern Siberia is two to three days. Swedish winters, too, are becoming shorter and less snowy, though the region of the world that has lost the most snow cover since 2000 is South America, at more than 20 per cent.

The albedo of the snow can be reduced by dirt and by particles that travel long distances in the atmosphere and land in snowy areas. The particles can come from a variety of sources: forest fires, agriculture, livestock, industry, construction projects, traffic. The dirt and particles accelerate the melting of snow in spring, when the sun is higher than in the winter, leading to faster evaporation. Warming due to lower snow albedo also increases the risk of droughts and leads to more frequent floods, due to earlier and faster run-off, which lowers the remaining water levels and can affect the production of electrical energy.

The disappearing snow also means that animal species that have evolved to be camouflaged for a long snow season fall prey to predators more easily. Less snow makes plants and animals more vulnerable to low temperatures, as they lack the protection the snow cover affords. Changes to entire ecosystems in snowy regions are already under way. Mammoths and woolly rhinos disappeared long ago, as did Europe's giant moose, and now many more species are teetering on the edge. Arctic fox, lynx and wolf. And the snowy owl, with its angelic white wings.

In Scotland, concerns were raised by a unique event: the first time it was reported that no snow at all had remained in the Scottish Highlands by the end of the summer season – in 1933, during what would turn out to be a strangely warm decade for the North Atlantic. The small area of snow in the Western Highlands known as the Sphinx, which always used to linger through to autumn in its north-easterly position on the mountain of Braeriach, sheltered by a high cliff facing the Atlantic, melted away completely. That was the first sign. Maybe it would be temporary? Most people believed it would be, and by the following summer, everything was back to normal again, with a snowpack that continued intact into the next winter.

And so it remained for another twenty-six years until, in 1959, the Sphinx disappeared once again. That, too, could be explained away: it was a warm and unmatched year for wine production in Germany and France, and it was also warm in the British Isles. Twenty-seven more years passed before the next full melt, in 1996, then it became noticeably more common, with total melts in 2003, 2006, 2017, 2018 and 2022. In 2019 the last bit of the snowpack made it through the year by the skin of its teeth; 2020 was also tight, the snow only making it through the summer because the snow cover had been exceptionally thick, with twice the normal snowfall in the previous February. Of the twenty-four years from 1996 to 2019, half have had one, two or no patches at all survive the summer. That change is nothing short of dramatic.

It is anticipated that the complete disappearance of Scotland's snow by the end of its summers will become the normal annual state of affairs as the global average temperature increases further. One model shows that by 2080 there is likely to be no snow in the Highlands even during the winter. Any snow that does fall might have a small chance of surviving the summer if there's been an unusually large amount during the winter, or if the spring, summer and autumn are colder than usual, or if next year's snow comes early and is plentiful. In other words, under exceptional conditions. Not so surprising.

But these kinds of conditions are increasingly rare. In 2020 there were abnormally continuous strong, south-westerly winter winds that brought unusually large amounts of snow to the leeward side of north-east Scotland, where the summer sun rarely hits. The words of the refrain in François Villon's famous *Ballade des dames du temps jadis* ('Ballad of the Ladies of Bygone Days', probably written in 1458), '*Où sont les neiges d'antant?*' – 'Where is the snow that fell last year?' – are increasingly relevant in Scotland, in ways such ballads could never have imagined. The only thing that could seriously disrupt these predictions would be a sharp increase in winter precipitation, itself attributable to climate change. Such scenarios are part of the models, including for Scandinavia, where snowfall in the mountains is expected to increase. But this is in contrast to what research is increasingly showing: that it's rising temperatures in spring, summer and autumn that will determine the long-term future of snow and glaciers, not the amount of winter precipitation.

On 19 August every year there is an 'All-Scotland' snow survey where large numbers of amateurs and enthusiasts are mobilized to go out during the summer holidays and search for any snow left on the ground. On this particular date in 2020, they found 179 snow patches, an unusually large number. By the time new snow arrived in November, there were twelve left, three of which were on Braeriach, the snowiest place in the whole of the UK, with three more at Ben Nevis, the UK's highest mountain, whose peak is 1,345 metres above sea level. Making the observations on the same day each year is important. It allows comparisons over time to be made easily and convincingly, while creating both a festive atmosphere and sense of civic responsibility, making the snow a point of concern even for those who might not otherwise think much about it. A form of civic education and an act of reverence for something that's in danger of disappearing.

Snow lingers longest in a place called Aonach Beag, which neighbours Ben Nevis close to the Atlantic coast, where low-pressure weather systems bring heavy precipitation during winter.

Snow-patch-hunting began in Scotland among mountaineers in the nineteenth century. Members of the Scottish Mountaineering Club made it a hobby but also earned goodwill through their efforts out in the wilderness. They contributed to the knowledge of their country, creating new data that people could add to what they already had on the changing seasons down in the valleys, and on the rhythms of farming and time. Scientific researchers followed their example from the 1940s, when they started their own snow-field studies, and since the 1990s the surveys have been conducted by recreational researchers, again drawing their ranks from climbers, skiers and other outdoor enthusiasts. Each year the Royal Meteorological Society publishes a report featuring the results that have been submitted to them. The report repeatedly shows there can be no doubt that the disappearance of snow is an effect of increasing temperatures.

In 2021, one of these devoted amateurs, Iain Cameron, wrote a book about his hunt for snow, *The Vanishing Ice*. The title is strange, given that the book is mostly about snow. In it, Cameron describes summer snow as a kind of relic, which in many cases has survived, year after year, since the Middle

Ages, until it vanished in 1933. A similar melting might have occurred in the seventeenth century, but there was no monitoring of snowfields then, so it remains conjecture based on historical climate calculations. Cameron's book is full of accounts of him wandering the gentle Scottish Highlands, where the peaks rarely rise more than 1,200 metres above sea level. In Cameron, too, we see an emotional aspect to his relationship with the snow he maps so meticulously. There's no contradiction between numbers and emotions, he holds. Quite the contrary. The more data he and his snow-patch-hunting colleagues collect, the deeper his feelings of loss and grief grow.

This kind of emotion requires expression, and in his book Cameron talks about the way the terminology we use has shifted; how insidious that is – and how short our memory is. When the snow first disappeared, in the summer of 1933, members of the Scottish Mountaineering Club spoke to *The Times*, declaring that they were convinced such a thing would never happen again on Garbh Choire Mòr, the hollow on Braeriach where the Sphinx may be found. From then on, they said, the snow would always stay put. At the time, Cameron tells us, the term 'perpetual snow' was in use. Then the word was changed to 'semi-permanent', and later to 'semi-perennial'. Cameron assures us that he's not an academic; he just sees what he sees, and he has read the diaries of those who walked the Highlands before him, counting the number of snow-patches. There's no doubt about it, he says: it's moving away from him, despite the assertions of his predecessors. His book is a beautiful piece about working through the grief he has experienced during his prolonged encounter with a landscape whose snow he sees disappearing before his eyes. It's 'the idea of Scotland' that's disappearing, he writes. It's about no longer being able to see the north face of Ben Nevis from Càrn Mòr Dearg in the July light, with its ravines, hundreds of metres long, filled with untouched white snow. 'How sad it will be for all of us to be deprived of such views.'

In the Sierra Nevada mountain range in southern Spain, a middle-aged man walks along holding an ice-pick. Jose Antonio Peña helps the water to find its way down into the arid landscape in the hottest corner of Europe, a role he's inherited from his older predecessors. He doesn't know if anyone will take over the job from him. He feels like the last person in a thousand-year history of managing the annual snow-melt. This is one of the most important functions of snow, which it performs all over the world: seasonal water storage. The water is released when spring comes – or when the next melt occurs, since they are happening more and more frequently, often in the same year. In the arid provinces of Andalusia where the waters of the Sierra Nevada eventually end up, snow from the mountains has been the primary source of water since the eighth century, with the colonization by the Moors in the early Middle Ages.

The vast network of *acequias*, as they're called in Spanish, comprises three thousand kilometres of hand-dug channels and streams that discreetly branch out into the landscape. In Arabic, the word was *as-saqiya* – conductor or carrier of water. These man-made waterways slow the water's journey down the mountainsides. Instead of rushing into rivers and disappearing into the valleys below, the water slowly seeps into the soil along the channels, in which it accumulates. The water is dispensed in a kind of aquatic democracy, giving life to all the vegetation of the area, which is how crops can be grown – and the farmers can enjoy more stable conditions. Vegetation, in turn, binds the water to the site. This makes the landscape more verdant and pleasant. This method has been able to support agriculture and viticulture for more than a thousand years. The first seven hundred years were dominated by the Moors, who introduced a wide range of new crops that required water: almonds, artichokes, chickpeas, aubergines, pomegranates, spinach, watermelons. The latter five hundred years were under Christian hegemony. All the while using the same basic technique and life philosophy, surviving the heat and averting drought with the help of the snow.

For some decades now, however, this has been increasingly difficult. A growing number of people from the region, especially young people, move to far-off cities for their education and work. Which means water managers like Peña have few successors. Throughout the twentieth century, Spain's intense pursuit of more water pointed in a very different direction. While the *acequias* were left to their fate, the country's political elite, in collaboration with the military and engineers, carried out a series of large-scale municipal projects such as dams, canals and tunnels, which involved the forced displacement of people and, under Franco's regime, the forced labour of political prisoners. This enabled urbanization, international tourism and industrial expansion, but at a high price, both ecologically and socially. Spain today is vulnerable to climate change and plagued by major regional tensions.

These are challenges that can still be addressed politically. Another challenge is more difficult: the snow. It no longer falls as reliably and regularly as it used to, and when it does, it isn't as abundant, even in the mountains. In the Middle Ages, the Moors dug storage places, *pozos*, cave-like stone chambers high up in the Sierra Nevada, where the snow could be stored and protected from the repeated melts. The snow was packed and could be made into different-shaped blocks of firn (old, compacted snow) and ice that were then transported down to towns and villages in summer by snow traders known as *neveros*. In the early 1900s, the city of Granada alone received a ton of such ice per day for use in healthcare, hotels, restaurants and affluent households. Similar methods have been used all around the Mediterranean and also in North Africa.

This is long in the past now. The *neveros* have been replaced by artificial ice-making. Winters in the region have also become milder and drier, and the amount of residual snow has decreased. Shorter snow seasons have diminished the amount of snow that can be usable as water in spring and early summer, the critical period for the irrigation that will supply the entire summer crop. This is true in Andalusia, but it is equally true in other parts of the world. The older method reduces vulnerability and makes the distribution of water more equitable, without much infrastructure. But it does require snow. Snow is the natural form of storage for water on steep terrain with few lakes, especially in hot climates with high evaporation. Which means snow can make all the difference as to whether or not a place is even habitable.

Light and Silence

*

THE TELEPHONE RINGS AT SIX O'CLOCK in the morning, something we've become used to. 'This is a message from Princeton public schools,' a booming voice intones. The first time we got a call like this back in November, I expected to hear that a nuclear war had broken out. But no, it was what they call a 'snow day'.

I'm living in New Jersey for a year as a fellow at the Institute for Advanced Study in Princeton, whose address – I Einstein Drive – is named for Albert Einstein, who was the first member of staff to be hired. We are neighbours with the large university in Princeton. My youngest daughter has come with me, and it's her school that's contacting us. On a snow day, the schools are closed 'due to inclement weather'. We have a similar expression in Swedish, too, *otjänlig väder-lek*, but few people use it, because our modern hubris means that we no longer anticipate having to adapt to changes in the weather.

On this February day, it's been snowing heavily for more than 24 hours. Dense, heavy snow, falling at sub-zero temperatures. During the night, warmer air has moved into the higher atmospheric layers, turning the snow to rain. It's called 'freezing rain' – raindrops that immediately freeze as they meet the cold ground, the snow, and all the trees and telephone wires, which they still have above ground, hanging, dangling in the air, as though we were in India and not right in the middle of one of the wealthiest regions on the planet.

We're already awake by the time the phone rings, awoken before dawn by loud banging sounds like rifle shots from the Institute's gardens. This is the sound of trees breaking. Mighty branches of the cedars in the Institute's grounds come crashing down, broken by snow encased in just this kind of frozen rain. My daughter and I watch the spectacle in amazement. Not just a few branches, but lots of them. Hundreds of pounds each. They fall one after the other, as though they were leaves dropping in autumn.

There's nothing to be done. Just as in the case of the sudden freezing in Rondane. Once the atmosphere has decided to take action, there's nothing we can do to stop it. Our only chance is to get out, ahead of it.

One of the best-known large storms of this kind hit parts of Quebec, Ontario and New England in January 1998, dropping frozen rain for eighty hours straight, knocking out power systems and snapping trees over an area the size of Sweden. The whole city of Montreal was enclosed in ice. The Canadian military deployed the largest force mobilized since the Korean War, sixteen thousand men, none of whom could do anything other than to help clean up afterwards.

My daughter and I walk through the almost laughably opulent residential neighbourhoods. The banging sounds continue, albeit less frequently. The traces of the night's slaughter are everywhere. Uprooted trees, bushes completely weighed down. We walk in the middle of the street to reduce the risk of getting hit by a falling ice-covered branch. Or a fallen wire, as in Ang Lee's suburban drama about betrayal, *The Ice Storm*, where an electric shock kills an innocent teenager in a world where most others are morally compromised.

Why are we even out in this state of emergency? I don't think it's due to some kind of death drive, but out of a deep, peculiar longing to make direct contact with the forces we're contending with. To feel Gaia tremble.

Over at the Institute, even the library is closed. The only thing open is the canteen, which is teeming with lunch guests. At one of the tables, I catch a glimpse of a very elderly John Nash, winner of the Nobel Prize in Economics (also known from the 2001 film *A Beautiful Mind*, in which he is played by Russell Crowe). The atmosphere is strangely upbeat.

The whole day feels cinematic – like something supernatural. A taste, I suppose, of what it might feel like to have survived the apocalypse. I can only remember one similar situation. It was on the *Lapplandspilen*, the 'Lappland Arrow' night train down to Stockholm from the north, in the 1980s. There'd been a delay and the train was at a standstill in the bright winter sun among snowy fields in Hälsingland. The day was already long past, and everyone had given up whatever plans and ambitions they might have had for that evening. The staff, in a bout of profound wisdom, had decided to give away all the stock from the restaurant car. It was a stationary, jovial party, with the sparkling snow and blue sky as witnesses. It's the only train trip from this time that I can remember in such vivid detail. I can only hope that it really happened.

Everything is still while snow falls. On their way to the ground, snowflakes spin in the face of the slightest wind.

In stormy weather, they move as much across the ground as towards it. Afterwards, the snow lies in whatever shape it was left in by the wind. Few things appear as restful and motionless as an expanse of freshly fallen snow. Or so we like to think. If we could only get really close, we would see how false this image is. The snow doesn't have time to reach the ground before the crystals begin their transformation – or you could call it its compression and erosion. Even in severe cold this process is ongoing, and after a few hours the airiness of the crystals has already fallen victim to gravity. If the temperature rises, the process accelerates and the snow becomes moist, heavier and even more compacted. If exposed to wind, snow can form ridges, edges, beds, drifts and so-called hanging drifts. A special case is *sastrugi,* from a Russian word for a furrow or small ridge. *Sastrugi* snow consists of ridges packed tight by the wind but with softer layers of snow that are easy to trample down into, which can make you fall.

Moist snow forms a crust on the surface layer, and the harder the crust gets, the better its resistance to further impact. However, given enough time, when the sun is highest in the sky it will penetrate even the hardest surface. Rain has the greatest melting effect, however, especially in really warm weather. Water penetrates the snow in a short time and dissolves it from all sides. Wind also has a pronounced effect. Wind enhances evaporation, which is known as snow sublimation. If the wind is also warm, the snow can disappear at a furious pace without any melting taking place, so no meltwater can be observed. Wind can create enormous excesses on the leeward side of a slope from quite a small amount of snow. Tons of it can be hanging around, just waiting to fall. 'Wind creates avalanches,' wrote Henri Bader, the Swiss snow physicist who spoke about snow's metamorphosis. Its continuous transformation.

Eventually, when snow is melting but hasn't yet disappeared completely, it gets stored as a kind of residual snow that stays around until the following year. This is what happens on glaciers. The precipitation from above is stored on the ice, while at the same time the glacier melts at the 'tongue' further down, where it's warmer. The very first glaciologists, in the early nineteenth century, were fascinated by the slow formation of glacier ice. The process was quite different from the formation of ice on water, which people had long been interested in. Few people had considered how glaciers form, and fewer still thought that they might be remnants of an ancient ice age, even if the ice itself has been replaced many times over. The biologist and geologist Louis Agassiz, who in 1840 wrote the very first major work on the Ice Age, imagining the ice to be many thousands of years old, was also one of the first people to start thinking about how glacier ice regenerates. He came to the realization that it had been through snow. During his walks among the Swiss glaciers, he noticed uneven, heavy grains of something that was neither snow nor ice, that appeared on top of the glacier during the summer. He realized that this was a transitional form that, during the cold winters, would eventually become solid glacial ice, under a combination of pressure and cold.

These round pellets of snow, commonly called *Firn* in German (as well as in Swedish and English), but also *Kornschnee* in Switzerland and *névé* in French, can be likened to coarse buckshot, having a diameter of three to five millimetres. There is plenty of space between the grains of firn for exposure to air and moisture, which will cause it to melt or evaporate. This, of course, created a mystery: how could the firn survive and form new ice under these kinds of conditions? During Robert Falcon Scott's ill-fated South Pole expedition on the *Terra Nova*, his fellow glaciologists Charles Seymour Wright and Raymond Edward Priestley observed that firn grains, in their 'evolution', if I may call it that, no longer exhibited a spherical shape. Between them they had formed a connected layer of ice. Wright and Priestley wrote, 'During the further growth from névé to ice, the air no longer forms the outer boundary of the crystals, but is included in them in the form of spherical or elliptical bubbles.' Their book on this was simply titled *Glaciology* and was published in 1922. By then it had been ten years since Scott had died in his tent during his trip back from the South Pole, in his inadequate clothing and with paltry provisions, but with a well-maintained diary. And snow – which Scott never became comfortable with – surrounding everything.

Wright and Priestley weren't the first to make this observation. Fifty years earlier, a Swiss geologist, Albert Heim, had referred to firn snow as 'white ice' – white because the firn was full of air – like the way ice looks when there's a crack in lake ice that then re-freezes. A precursor to ice. During winters, this proto-ice is compressed by fresh layers of snow that leave new firn on top of the old, which in its turn freezes into a more cohesive form of ice, with the bubbles still inside. The ice only has to survive its first few summers to be guaranteed a long life as a gradually descending layer of increasingly solid ice, which won't melt until it reaches the lower parts of the glacier. Depending on the size of the glacier, the slope of the terrain, the temperature, precipitation and the frictional properties of the underlying soil, this journey of historical air can take centuries or millennia. The Greenland Ice Sheet is more than 100,000 years old, while in Antarctica scientists have found ice that reaches 4.6 million years into the past.

Scientists in the early twentieth century also started measuring the density of ice. Firn, which contains more air, was the lightest. White ice was heavier, and fully formed ice heaviest because the air bubbles had been largely pressed out of it, though not completely. The German climatologist Alfred Wegener, best known for his theory of continental drift (that is, that today's continents were once connected in the primordial continent of Pangea and that they have been floating around the earth's surface for millions of years) had already in 1913 drilled twenty-five metres into the ice sheet in north-west Greenland with his colleague Johan Peter Koch, finding that there was still air in there even under ten standard atmospheres of pressure. What neither Heim, Wegener, nor Wright and Priestley paid much attention to was that the air inside the ice was the same air that had existed when the firn froze solid. Although it occurred to them, it wasn't something they attached any importance to. Others would come to do that. Others whose minds would slowly move in this direction, and suddenly see something truly novel.

A few decades after Wright and Priestley's observations – 1948 to be precise – physicists, chemists, biologists and physiologists gathered at the University of Copenhagen. They were seeking answers to new questions about the dating and timing of both life and the climate in the harsh conditions of the North Atlantic and Greenland. Life there was lived on the margins. But how had it actually evolved and what clues were there that might provide answers? Among the researchers were Nobel Prize winners Niels Bohr (Physics), August Krogh (Medicine) and George de Hevesy (Chemistry), the latter a Hungarian who spent many years in Copenhagen but who now was primarily working in Sweden. Still more people were experts on the radioactive isotopes that could give us more information about the past through organic material, in ways similar to the then-new carbon-14 dating method, but with meaningfully greater scope as regards the timescale and the kind of material used. The leading figure behind these meetings, which were also attended by Quaternary geologists (specializing in the relatively recent geological past, up to 2.5 million years ago), oceanographers and botanists, was the German-Jewish physicist Hilde Levi. She had just returned from spending time in Chicago, where intensive work on isotope dating was already underway.

Part of the contemporary context included increasing discussion regarding the colder climate in northern Europe during the late Middle Ages, and how this might have contributed to Norse settlers abandoning Greenland. The cooling that took place during the later Middle Ages onwards had been widely known for a long time. It was part of the 'Little Ice Age', a concept dating to 1939. However, by the late 1940s it had been given new relevance by archaeological and volcanological research taking place in Iceland. One important contribution was a summary of the latest expertise in the field by the Swedish historian Gustaf Utterström, published in 1955, in which he clearly favoured climate change in the Little Ice Age as the central explanation of the Norse migration from Greenland.

The debate about the importance of climate to the Norse in Greenland would return, but in a way that no one could have imagined, when the Copenhagen meetings began in the late 1940s. One of the main characters in this story was a brusque young physicist with wide-ranging interests and an unquenchable curiosity, who was investigating how precipitation might function as a repository for isotopes. The young physicist, a PhD student enamoured with history, was Willi Dansgaard – and he wanted to get his hands on 'old water'. The idea had originally come to him one rainy week in June 1952 as he was testing a newly acquired spectrometer at his home institute of the University of Copenhagen. With it, Dansgaard could see that multiple varieties of oxygen isotopes appeared in differing amounts in raindrops. Since the distribution of isotopes reflected the temperature of the water droplets, he could trace where in the atmosphere the rain had come from. The heavier ^{18}O isotope was more abundant in the colder droplets, when water had been formed higher up in the atmosphere, where the temperature was lower – a modest observation in itself, but one holding great potential. In practice, this meant that water, the most common substance on earth, could offer a peephole into the weather of the past and, perhaps, 'old climate'. Because, of course, carbon dioxide and other greenhouse gases were also preserved in the same water.

However, it soon became clear that the method was most useful when applied to ice, which was actually 'old snow'. So, it was to ice that Dansgaard now turned his attention. But where could he find this ice? Current geopolitical developments presented him with an unexpected opportunity. During the Second World War, the United States had started using Greenland as a base. Now, during the Cold War, America made even greater use of Greenland, building new bases and also embarking on extensive research into snow and ice (which we'll come back to). However, Greenland belonged to Denmark, so the US and Denmark needed to work together. In studying the Greenland Ice Sheet, they enlisted the help of Swiss scientists that the Americans had already started employing. Together they wanted to investigate what was hidden in the depths of the ice. This is how Willi Dansgaard was able to start working with long ice cores that could give him more detailed knowledge about the historical sorting of isotopes.

And the isotopes holding that important information were, in fact, in the air bubbles. The very same kind of air that Wright and Priestley had noticed in Antarctica when they studied how the ice was formed, and which they had seen frozen in between the firn balls. Raindrops had once offered intriguing questions and answers. But if, like Dansgaard, you were interested in climate as a historical factor, causing the Norse exodus from Greenland, this information, which ultimately came from snowflakes, would be more useful. It was, after all, snow crystals that turned into ice, and they retained much of their atmospheric labelling throughout the formation of the ice. A kind of microscopic message in a bottle from the time before written language.

Ordinary water that freezes in oceans, lakes and rivers can't carry air in the same way, because what freezes is just . . . water. And most importantly, that ice is annual and doesn't form repositories. Neither does that of the mountain glaciers, whose rapid movement downward rips apart the layers that form annually. Researchers needed to get right up to the top of the three-kilometre thick and very slow-moving Greenland Ice Sheet to unroll this whole panorama of snow and time. It was soon discovered that the snow in the drill cores stretched back thousands of years, even tens of thousands, even hundreds of thousands of years – or more.

During the 1960s and 1970s, Dansgaard worked intensively on the drill cores in Greenland. The analyses were gradually refined, and during the 1960s it became increasingly clear that the climate had changed dramatically throughout this long history – and several times over, in fact. And that such climate shifts could happen very quickly, within centuries or even decades. Over time, his own interest shifted more and more towards historical issues. One old question that could now be attacked using the new methods was precisely that of the decline of the Norse civilization in Greenland. Dansgaard tackled the issue with great eagerness. He was influenced by Utterström and many others who had debated climate issues from medieval times. And he was supported by the doyen of snow physics, the Japanese physicist Ukichiro Nakaya, who'd also spent time in Greenland in the 1950s talking about how the ice cores 'trapped the air inside the snow like air bubbles'.

So here was a Danish physicist trying to solve the riddle of what the weather had actually been like in Greenland during the late Middle Ages, when the last Norse had left the area. Who could have imagined that the life cycle of snow would give rise to this magical movement of time and information? An innocent snowflake travelling through the air, spinning, swirling, floating, seemingly not lasting very long – it was the properties of this very crystal that enabled the insight which today has presented us with perhaps our greatest challenge. Once that insight came into focus, our knowledge began a new journey, in the form of chronologies that could be compared to those already found in lake sediments, tree rings, bogs, clays and fossils. However, none of these elements could tell us as precisely and in as much detail about what climate had been like in the past as the ice cores could.

Snow had conquered the world, creating the basis of our modern understanding of it. The new history of the world presupposed the history of snow. And it did so using man-made, anthropogenic climate changes that were part of the history that intertwined humanity with the earth and all its spheres, not least air and water, the atmosphere and the hydrosphere.

What we're able to talk about now seems at once so far away and so close. It happened in my own lifetime, though I was still a small child during those years. The realization came slowly, but the change of light was sudden. Like dawn. Gerald Seligman, a British geographer and glaciologist, was at the very centre of this issue. In his significant book on snow and ice, *Snow Structure and Ski Fields*, from 1936, he is interested in the problem of how glacier ice forms. Seligman makes his way through the research, interested in two things: first, taxonomy, and second, which terms are best suited to denote different stages of the snow-to-ice transition process. I'll return to him a few times throughout this book. Seligman was a central figure in snow research, founding and leading professional societies, and launching and editing their publications. In the 1930s he took part in an expedition to Graham Land on the Antarctic Peninsula, and later became Dean of Trinity Hall, Cambridge, where he himself had studied. He was considered important enough to have a bay named after him, on Graham Land itself – Seligman Inlet. Seligman tried to instil a form of order in glaciology; to hold certain boundaries. The presence of carbon dioxide in the atmosphere had no part in that. When someone else suggested that this greenhouse gas might have something to do with glaciers, he held resolutely that this could not be the case.

At the same time, not far from Seligman, just north of London, there was an engineer, Guy Stewart Callendar, an expert on gases. Callendar was a marginal figure within his contemporary scientific community, working mostly alone, collecting temperature data at his kitchen table using newspapers and data from Kew Gardens, the botanical garden founded in the eighteenth century and then an imperial hub of London, where species and data from around the world are brought together to this day. How hot, he wondered, was it getting in different places around the world? It was evident to Callendar that it was definitely getting warmer, anyway. Callendar also measured the consumption of coal and oil, which was increasing. His article on this appeared in 1938 in the *Quarterly Journal of the Royal British Meteorological Society*. He was convinced that carbon dioxide in the atmosphere, and by extension humans, were responsible for climate change. Very few people believed him. Gerald Seligman's rebuttal was short and emphatic, on an issue he really didn't know very much about at all. Nobody knew anything. Snow and ice were considered unconnected to human activity for an oddly long, protracted period.

Snow has many properties that might not immediately spring to mind. For example, snow is a crystal, but while other crystalline materials like minerals are hard and melt only at very high temperatures, snow is soft and melts at the slightest touch, and at a temperature of 0°C no matter what. Snow has great sound-absorbing capacity; it is associated with silence. Freshly fallen snow, in particular, absorbs sound very effectively. When snow falls, sound waves are captured by the snowflakes. Each little hexagon makes its own contribution. When they land on the ground, this sound-absorbing property remains. Even five centimetres of snow in a landscape reduces sound by half.

Sound absorption decreases as the snow compacts, as this breaks down the finest structures of the snowflakes and squeezes out air trapped in the pockets that function as a sound-absorbing cushion. When the snow becomes more densely packed in late winter, and when there is thaw or rain that form snow-crusts if they then freeze, the sound-absorbing properties of the snow are almost entirely lost. The effect is increased by soil and dirt in the snow. And if the snow melts almost completely, and ice forms on the ground, sound bounces even more easily and quickly off that hard ice, especially if holes in the ground have been levelled out and vegetation is more barren. Then snow amplifies sound. Slippery and unromantic.

The silence of the snow is most noticeable during the dark part of winter, when there is fresh snow on the ground and dense falling snow in the air, especially *lapphandskar*. Then the sounds of the world are but shy murmurs. The silencing effect of snow is enhanced by cold. At low temperatures, air density increases and sound waves need more energy to travel.

The silence of snow is soothing. The amount of stress hormones in the body decreases. Psychologists and criminologists have long known that people tend to commit fewer serious crimes (robberies, assaults) when there is snowfall and snow cover on the ground, which appears to be supported by statistical analyses. Fewer crimes are committed in the cold and significantly more in intense heat (up to 30°C), an insight that was already known to the father of social statistics, the Belgian Adolphe Quetelet in the nineteenth century. However, higher winter temperatures also appear to increase the propensity for violent crime. The United States has reported several cases of abnormally high crime rates on warm winter days, and there are concerns that more volatile weather, which is thought to accompany climate change, will lead to more serious offences.

So snow, when aided by cold, seems to dampen hostility and aggression. The mood is anchored in our culture. The snowfall in romantic comedies signals relaxation. The search is over; harmony and union are what now await the couple. Forgiveness might be another term. Literary snowfalls have made use of the same meaning. Perhaps the fact that things are frozen contributes in some way. The landscape is frozen and embedded, making many actions superfluous – we think, akin to the hibernation of hedgehogs and bears, of rest. The farmers' days are now quiet. The forester, on the other hand, comes to life, as do the hunter and the skier. Children rush out – resting is the last thing on their minds. They're the source of sound. All around them, the forgiving silence. Maybe, too, the ultimate forgiveness. The snow and the snowfall also signal the end of life. Our final rest. When the cries of life are emitted and we are forever under the snow.

Snow is also the opposite of silence. Snow creates a multifaceted soundscape. There's creaking in cold weather. The crystals are extremely brittle and when they break there's friction. The colder it is, the brighter and shorter the sound. Some snow crunches like crispbread. A munching sound. The Nobel Prize winner in Chemistry at Umeå University, Frenchwoman Emmanuelle Charpentier, received her prize alongside Berkeley researcher Jennifer Doudna for the gene 'scissors' CRISPR-Cas9. When Charpentier came to Umeå in the far north of Sweden to do research and she set down her shoes in the snow for the first time, she thought that every step she took sounded just like that: *crispr, crispr*. 'When I heard it, I said, "That's a sign."' Still, other kinds of snow whisper like the softest silk. Snow melting and dripping from the eaves is like synthesizer music. Snow-crust sounds are looking for attention. Old snow sounds more open than new snow, which tends to muffle noise.

The real decibel-extreme is a blizzard – and these can be deadly. The worst blizzard catastrophe on Swedish soil in modern history took place in the Anaris mountains in February 1978. Short days. The temperature -16°C down to -20°C. Thin snow cover. A party of nine set off, with little mountain experience between them. A sudden weather-change: hard, icy, downslope winds. The roof of the emergency bivouac blows off. Snowdrifts several metres high. Zero visibility. Three days. Eight people die. The ninth has to have his feet and fingers amputated. What must that roaring have sounded like?

Anyone who has experienced a snowstorm without the possibility of taking shelter can testify to the sheer terror of it. And to the roaring of the wind through the mountain birch forest, the waves of which can be glimpsed through the milky grey-out of the snow. The rest of Creation ceases to exist. Like being in a frozen cellar, not knowing if you're going to emerge alive. Even outside of the mountains, severe snow storms can occur. On 29 January 1850, a sudden and violent snowstorm hit Sweden's north-eastern region of Götaland and eastern region of Svealand not far from Stockholm. Around a hundred people are thought to have frozen to death. The so-called 'Snowstorm Tuesday' is one of the deadliest single events of any type to have taken place in Sweden. What must that snowstorm have sounded like?

The sound of a storm without trees was recorded in a blizzard report from Spitsbergen in Norway in 1934, when the Austrian artist Christiane Ritter, aged thirty-seven, let herself be convinced by her scientist husband to spend a winter in a small cabin at Gråhuken, by a bay facing the Arctic Ocean. It was his fourth winter there and her first – and the first by any woman at all, that far north. What she encounters is another world, at once terrifying and marvellous. And she finds herself confronting forces she doesn't recognize or understand. She describes them in her book *Eine Frau erlebt die Polarnacht* ('A Woman in the Polar Night'), which was later translated into several languages, selling more than half a million copies by 2012.

On one occasion, when her husband is off hunting with a friend, a storm blows up that is like nothing she's ever seen on the southern European continent she's left behind. During the nine days and nights of the blizzard, she records a whole series of monstrous sounds emanating from the living entity that a massive storm comprises, which Ritter refers to alternately as a monster and a deity. The storm is an 'express train running uninterrupted over iron bridges and through endless shrieking tunnels'. The men are supposed to be gone for thirteen days, but have told her 'don't worry if we're away longer'. She goes outside only to find the waves of the sea crashing high up in the air off the rocks. But the storm itself is louder than even the sea, like a 'deep, drawn-out organ note'. The tiny amount of heat given off by the stove 'whistles' from the chimney. There's also whistling through the cracks in the walls. After a few days, the 'hollow rumbling' of the still-raging storm transforms into an 'incessant roar' accompanied by the crashing of waves against the rocky shore outside. An attempt to sleep fails, not due to any lack of fatigue, but because of the pounding and scraping of the loose boards against the east wall of the cabin, along with the banging and clattering of the now-frozen bodies of foxes that had been shot, skinned and tied to the roof. The storm was still rising, with one sound soon indistinguishable from the next. 'Everything came together in a numbing roar,' she recounts.

Ritter crawls out of the ten-metre snowdrift the storm has built in front of her cabin to fetch coal and paraffin, but as she emerges from this snow edifice she finds herself unable to move, due to the force of the wind pressing against her body. The bellowing deity is not only an all-encompassing noise, it's also an immutable, all-encompassing wall. And this absolute, total, relentless noise continues for many more days and nights – what Ritter ultimately refers to as 'the fury of insane music'. Her cabin shakes, and by the end she starts shaking and trembling, too, as if she had become one with the ferocity of the elements during the nine-day blizzard.

Christiane Ritter's final words in her book could suit our own time a century later: 'The vengeful gods return. Morality reawakens, throwing itself at humanity like a monster.' As though this elemental fury requires interpretation. And the only thought that ultimately remains an option is that the gods are angry with us. Because this rage is evermore something of our own creation. Ritter's depiction of her storm calls to mind Joseph Conrad's 1902 novella *Typhoon*, the archetypal example of the storm as something 'spectacularly sublime'. Conrad's story shows what happens when you try to make decisions, to navigate a ship without regard to the actual circumstances and without seeing people's ability to work together. The deepest insights are rarely new.

A katabatic wind is a downdraft (ancient Greek *kata* = down) formed by cold air on a plateau or glacier. The colder the air, the heavier it gets, causing it to flow down towards lower terrain, especially at precipices and ravines. Katabatic winds can reach speeds of up to 100 metres per second. They're common in Antarctica, while in Greenland a katabatic wind is known as a *piteraq* – 'the one that attacks you'. It's a katabatic wind that holds the Swedish wind record, 81 metres per second, which was measured in December of 1992 at Tarfala, a research station just below the mountain of Kebnekaise, the highest point in Sweden. Winds of this sort of strength drive snow, ice and gravel violently across the ground, clearing large areas while creating a terrifying noise.

Melbourne-based sound artist Philip Samartzis has recorded the sounds of katabatic wind. Inspired by early Antarctic photographers such as Herbert Ponting and Frank Hurley, he travelled to Antarctica first in 2010 and again in 2016, when he stayed at Australia's Casey Station. He took as his starting point two photographs in particular, both taken by Hurley in 1912 on an expedition by fellow Australian

Douglas Mawson, 'A Blizzard' and 'Leaning on the Wind'. A katabatic wind can be of such strength and stability that a human being can 'hang' in the wind without assistance. Samartzis's recordings try to capture this frenzied combination of wind, ice and snow. At Casey Station in particular, the conditions are optimal for the fiercest possible winds, with a great contrast between the icy, 1,400-metre high Law Dome mountain above the station and relatively mild conditions at the coastal station itself. The acoustic effects are remarkable. Sound travels all around. Wind can mask sound; the absence of wind can emphasize sound.

Equally evocative are the associations with the optical properties of snow. Where there's snow, there's light, even though snow is most prevalent during the year's darkest seasons. When there's snow on the ground, usually all it takes is a few lumens from the sky to provide enough light on the ground for you to easily find your way. Together, moonlight and midwinter snow can make the landscape vibrate as though we're witnessing the earth's own version of the Northern Lights. Moonlight and crusted snow enhance the effect somewhat, while fluffy new snow absorbs light as effectively as it swallows sound. The shiny layer of frozen moisture in the skies of late winter and early spring reflects even the slightest amount of light from the sky in a powerful vertical beam. So bright is the moonlight on the snow-crust of the night that trees and other shapes in the landscape cast deep shadows with sharp contours.

In the middle of the day, by contrast, the light is so strong that it is hard to see anything at all. One of the paradoxes of snow. If the sun is high in a cloudless sky and the snow cover itself is clean and continuous, not overshadowed by forests or buildings, the light is brighter than in a treeless desert, and frequencies in the dangerous ultraviolet spectrum are also amplified by their reflection off the snow. In overcast weather, the intensity may appear lower, but this is often an optical illusion. Moisture and particles in the atmosphere now act like prisms, scattering light and mixing it with reflections and other effects. On cold winter days when the air is filled with microscopic snow crystals, there can be halo phenomena in the sky, also known as solar rings. Beautiful to see, but they also have unusually bright light, and even in cloudy weather you have to squint due to the onslaught of light from all sides. It's not surprising, then, that since ancient times Arctic peoples have made snow goggles from driftwood, seal bones, walrus tusks, reindeer and caribou antlers, as well as beach grass. The openings to see through are very narrow, and the goggles are often intricately decorated, being greatly valued as vital for survival.

Under these conditions, in terrain having few features or vegetation, different forms of whiteness tend to merge and the contours of the landscape itself can disappear. This is what's called a whiteout. While in indigenous Nordic languages there are much older words for the same event, contemporary terms emerged in Western languages when flying in the polar regions became more common. In 1946, the word was used as a technical term by meteorologist Leonard Hedine from the US Weather Bureau, who wrote in the *Bulletin of the American Meteorological Society* about the need for a 'term in common use and understood by all' to describe the weather phenomenon he'd observed in Alaska. His proposal was 'Arctic whiteout', and it was soon employed to warn pilots of the danger in not being able to clearly make out the distance between the sky and the ground, nor being able to judge distances.

Another US Weather Bureau employee, Arnold Court, read the article and commented that he had heard the same phenomenon described by the term 'milky weather'. The term harked back to US polar explorer and pilot Richard E. Byrd's account of his and Norwegian-born Bernt Balchen's iconic first flight (along with a telegraph operator and a photographer) to the South Pole in 1929 – it was 'like flying in a world turned to milk'. For Hedine, it was important to distinguish between the newly coined weather phenomenon and the other already well-known terms. 'Milky', for example, was not the same as foggy. Nor was a whiteout the same as fog, snow blindness, sleet, a blizzard, or anything else. Hedine himself called the term 'descriptive'. What it's referring to is an 'extension of whiteness', he wrote.

The literary scholar Sabine Frost, whose conceptual enquiry I'm following here, compares it to the white areas on the page of a book. The old name for this is *carta pura*, meaning 'pure paper' in Latin. A 'whitespace' is the word used in English for a blank space between characters, words or lines – or it's termed a 'blank', which also means white. Similar words exist in several languages; *net vir blankes* was a common message in Afrikaans during South Africa's apartheid, meaning 'for whites only'. In Swedish we say *blankrad* for an entire line without text. The writing provides the contrast necessary for meaning to be created, and also allows the reader to distinguish up from down. A whiteout causes this textless whiteness to suddenly expand and take up more space.

A classic 'whiteout', as experienced by a polar pilot or someone travelling over snow on the ground, is characterized by a disappearance in the contours of the landscape. It is not the light alone that produces this effect; it is light in combination with snow. Primarily, it's the line of the horizon that vanishes, so that there's no visible boundary between earth and sky. At close range, the eye doesn't perceive the ground, or more accurately, the *snow*. You can see your physical hand in front of you, your shoes, your skis and your boots. But you don't see the snow that all of it's on, or in. Nor do you perceive, at least not with your eyes, the terrain ahead of you, whether that's an uphill slope, a plateau or a cliff. You can be standing on a precipice without having any idea it's there, other than from the sounds of the landscape, which change in the absence of having an acoustic base.

Even small changes in the landscape make it easy to fall. In rare cases, there may be a loss of kinaesthesia – that is, the ability to perceive position and movement – accompanied by confusion, physical imbalance and a general loss of functional aptitude. Like a book page without text, the reader no longer understands where they're going. In unfamiliar terrain, or in an aircraft travelling at hundreds of kilometres per hour, this is a life-threatening situation. A person in a whiteout is in the centre of the light, but this is tantamount to being in as much darkness, as if they were blind. But because there's light, you don't realize that's what's happening. That's where the danger lies.

Most of the earth's surface is water, with well over half of the world's population living near oceans, lakes or rivers. One thing we all have in common is that we can't see what's below the surface. It's often said that we know more about what's out in space, which we'll never touch, than we do about the sea, even though it's so close. Maybe that's why water both draws and frightens us – because we can't see what lies within it.

Perhaps we need to start saying the same thing about snow. The only difference is that far fewer people have ever found themselves truly close to it. The attraction is the beauty and romance that our culture has invested in snow, but to actually interact with it requires experience.

Maybe we need to return to the original experiences of childhood, when it comes to snow. We don't see everything that's happening. That means we don't understand everything about it, either, and when we don't understand something, we need to be even more alert, and keep gaining insight. In recent years, historians have started to include the oceans in their historiography. Others have recognized the place of the oceans in literature, art and film; *The Old Man and the Sea* has been filmed four times (and earned Ernest Hemingway both a Pulitzer Prize and a Nobel Prize), and *Moby Dick* even more. During our current climate crisis, Melville's book has also attracted attention for predicting an Earth totally flooded by a new deluge. Water, both the vast blue of the oceans and all that fills lakes, rivers and the earth itself, has in recent decades become the subject of our interactions with the planet.

Rachel Carson, best known for the sixties classic *Silent Spring*, about environmental toxins and how they ultimately bring about the end of birdsong, was a trained marine biologist and had spent most of her life writing about the oceans and the life within them. *The Sea Around Us*, published in 1951, was an eye-opening bestseller. Beneath the opaque surface of the oceans lay a miraculous diversity of life. What Rachel Carson did was to literally write out this abundance, making the richness of the deep visible. Only those who see it can have feelings about it, only those who feel then want to engage, and only those who want to engage can take action. Just as we once emerged from the oceans, Rachel Carson imagined, sooner or later, we'll have to return there, to take care of all the living things we hold together.

In the same way, I think, we need to get back to the snow. Into the vast whiteness. The silent, great and terrible colour that can save us.

Rehearsal.

On the night of 10 March 1883, Salomon August Andrée, a young, ambitious Swedish engineer and physicist, is at Cape Thordsen research station in Svalbard, Norway, where he's making meteorological observations during the ongoing International Polar Year. The strange thing he notices isn't the cold, which is the most common condition for the area; nor the fact that there's no wind, although that is rare. The strange thing is the total silence and stillness. No Northern Lights, which otherwise are out every cloudless night at these latitudes. No sound of ice roaring from the fjord below. Everything has stopped. Not even the electrical instruments are giving off the slightest signal. The condensation hygrometer, which indicates the humidity of the air, shows no precipitation. The magnetic readings remain the same, hour after hour. Not the movement of a bird's wing, not a paw in the snow. Life itself is missing.

Snow. A stillness without limits.

It was the white darkness of death that Andrée was

feeling. 'I almost felt horrible . . . I thought about how I would behave if I were here alone now, deprived of everything, and I found only one way out – the sea.'

Little did he know that fourteen years later, in October 1897, he'd be in the same situation again, this time on White Island after a walk across the ice. *The Eagle*, the hydrogen balloon with which he and his expedition partners hoped to reach the North Pole, had crashed there shortly after departure from Svalbard. This would be his last time on the island. His rifle rested in his hand as the sea lay before him. On that March night, he'd rehearsed his farewell to life, his escape from the loneliness and the cold treatment he'd received from the living, a fate worse than death.

Andrée's diaries include his thirteen-month period at Cape Thordsen as part of the First International Polar Year, of 1882–1883. The main aim of their stay was to take scientific measurements. These occupy a large part of his diary, interspersed with detailed accounts of the order of the meals, the many celebration days and the strenuous bouts of drinking at every special occasion – for name days and birthdays of royalty, with every guest, or for no reason whatsoever. Every now and then the emotional, reflective man breaks through. Andrée is at this point usually alone, having left the so-called *Svenskehuset* (the Swedish House), with its pressured and disgruntled atmosphere. He moves freely in the terrain with his rifle on his back. What he encounters is nature with its full potential: the winds and the cold, seals and reindeer, long-tailed ducks, eiders, sometimes a polar bear.

From time to time he fires a shot, as if testing the relationship between civilization, there only as a guest, and the permanent fixture of nature. The shot only seems to confirm that civilization is subordinate. No avalanche ensues. Here humans are small, and make themselves even smaller with their petty feuds and fleeting passions.

It's the same post-Darwinian attitude that can be found in the most developed Nordic art and literature of the era: August Strindberg, Edvard Munch, Karl Nordström. The parallels are perhaps most evident in Harald Sohlberg's paintings of the moonlit Rondane. Those two arched shapes in the snow. The divine presence of something 'else', above and beyond the icy valley of the shadow of death.

Snow Grief

*

I TAKE THE KICK-SLED TO SCHOOL from my house. I understand that if I'm going to succeed there I have to learn my vocabulary and file down my heron – the slender-necked wooden bird that my even scrawnier woodwork teacher, from the southern region of Skåne, cuts out with the band-saw for me to complete in the school basement with a rasp, file and steel-bristle brush. Ultimately, the fire-blackened heron will take its place in my house, as have those of all other pupils, atop the television, the central piece of furniture in the Swedish *folkhem* (as the welfare state was then fondly referred to).

The TV-heron and vocabulary learning are my personal responsibility. The snow – which falls one night from the sky – is someone else's responsibility. The fact that it falls exactly where I am is an act of grace. I don't use the word 'grace' very often as a ten-year-old. But now I do. Because it is indeed an act of grace. Or it was, because it's not really that way any more. More and more, we humans are doing what before could only be done by heaven and earth.

And we haven't thought through what's being lost. That which we can't replace. Species are disappearing.

Rivers are disappearing, turning into electricity. The silence . . . ceases.

The snow is also going to disappear.

Loss is intrinsically human. Grief and loss are timeless emotions. These feelings are particularly strong in the face of the most personal and private losses. Grief can be an almost unbearable state. For someone grieving, sometimes even relief through death isn't out of the question. In some circumstances, grief is also collective, with the people grieving together forming a kind of community that its members may not have been fully aware of before their shared grief experience.

There has been no shortage of losses to mourn. Wars. Sporting victories that didn't happen, leaving the home team in tears. Pride and honour, squandered through mismanagement and fiasco. Historical grief could have a substantial range. When a king or pope died, the rituals of mourning could be enacted, with some authenticity, at a great distance from Rome or the capitals, despite the long time it took for the news to reach them.

One strong motivation for collective mourning has been to nationalize the masses. In more recent times, the domains of grief have expanded considerably further and the pace of mourning has accelerated. Mourning now extends across the planet, covering the totality of it. The whole of humanity and its failures to live in peace with itself and with the earth have become the objects of our grief and despair. People have always had reason to complain, as Jeremiah once did about the state of the Kingdom of Judah. There was lamentation and weeping throughout the classical Greek tragedies. But even if lamentation could extend across whole kingdoms, city-states or even empires, there has never before been worldwide grief and despair. That phenomenon belongs to our own time. It was in the twentieth century that we seriously learned to mourn on behalf of the whole world. First on behalf of the World War traumas, and in the second half of the century most particularly, through the realization of our climatic and environmental vulnerabilities. Global grief has been of about the same duration as our modern appreciation of snow. Now they affect each other.

When looking for insight into what happens in these kinds of circumstances, one author on the subject is Judith Schalansky. In her 2018 book *Verzeichnis einiger Verluste* ('An Inventory of Losses'), she searches for the sentiment that makes our losses comprehensible. She visits places where these losses have happened. She describes them in detail, almost pedantically so. The sense of loss is palpable, but what has been lost is rarely clear. This creates a romantic undertone reminiscent of the preoccupation with ruined landscapes of bygone times, which in turn creates the impetus for preservation and documentation. Other losses are misunderstandings and myths that have failed to survive a more sceptical future. For Schalansky, it is important to recognize that not everything can be preserved. 'Living means experiencing loss.'

Maybe it's the resignation and powerlessness that the book can be said to represent that's made it so widely discussed. For its extended index of loss during our lifetime, its global chain of causation and diffuse, silent spread – like a powerful, chaotic virus – it's something of a handbook. But there is nothing in the book that is really about the transformations caused by climate change or the subtle, faceless terror of species extinction.

We miss the snow in more or less the same way. Vaguely, without very tangible arguments in snow's favour. We miss it in an individual way, though snow isn't an individual thing at all but, like the wind and the external temperature, one of the most common things in much of the world. And it's something we long believed was completely outside our influence – something many people still believe. A white act of grace.

Emotions slowly make their way towards our core, consuming us more and more. Sadness, fear, anger, despair, anxiety. Sometimes just an odd feeling that something is off. A nagging worry that something terrible could happen at any moment. Something uncanny, like a coming, all-consuming storm for which no official warnings have been issued, but which nevertheless, judging by all the signs, seems to be on its way. Shouldn't we feel sad when snow is vanishing around us? What would we call that kind of sorrow? Grief for the snow?

The evolutionary explanation for why we grieve is that grief teaches us to take care of our fellow human beings. If we were indifferent to the death of relatives and close friends, we wouldn't care as much about each other while we're alive. In so doing we increase our own chances of being thought about and cared for while we're still alive. This isn't to say that grief is inherently healthy. Quite the opposite; we don't voluntarily put ourselves in states of grief or despair. The loss of people close to us can mean that our quality of life deteriorates significantly. In the most extreme case, the loss of people whose presence and affection make life materially liveable and spiritually vibrant can rob a person of the entire meaning of their life. Risk of illness, even mortality, increases during severe bouts of grief.

None of this should be too difficult to understand. Most people have felt deep sadness, or can at least expect that they are going to do so. The same goes for loss. We usually think that these feelings arise from losing other people. It's people who grieve, we say, and that grief arises from the loss of another human being. And it's true, of course, that we grieve because these are people we care about. It's because of our fellow human beings that we live and survive in a meaningful way. The sole remaining human being on Earth wouldn't have this meaning. But does it have to stop there? Anyone with a pet knows that the loss of that animal can be as profound as the loss of a friend. So maybe that's what an animal is: a friend.

Animals also feel sadness. Darwin made this observation back in his day. Pliny noted that elephants grieve, and we know now that grief is something that many species experience. 'Evolutionary thanatology' – the scientific study of how organisms deal with death – is an established field of research into humans, as well as other species.

We can have profound feelings when facing the death or disappearance of nature, too. For instance, we wouldn't be surprised to hear that people with beautiful trees on their property mourn those trees when they're downed in a storm. And it is even worse if trees are lost due to cynicism, a behaviour that is unique to humans. When the 'Sycamore Gap' tree that had been growing in a dip in the landscape along Hadrian's Wall, on the border between Scotland and England, was cut down by vandals in the summer of 2024, it created a national outcry. A police investigation was launched and forensics officers were sent to take measurements and samples from the remains. One was heard saying: 'In thirty-one years of forensics, I've never examined a tree.' Entire land-scapes can be missed and mourned in the same way.

Why do we have the capacity to feel this way? It's simply because we've learned that it can be advantageous to have warm feelings for pets and trees that are near to us. It's simi-lar to the way we internalize social norms and feelings about right and wrong. There's also research that suggests the roots of this go even deeper, intertwining our collective learning with the age-old power of genetics. In the 1980s, entomologist and author Edward O. Wilson suggested that humans had an evolution-based capacity for 'biophilia' – that is, an 'innate ten-dency to focus on life and lifelike processes'. He wrote a book with this very title, *Biophilia*, which literally means 'friendship with life', a sympathy with the other living beings that surround us. Wilson conceived that there was something in the most minute structures of life that created an affinity transcend-ing the boundaries of species. Part of what constitutes being *human* is recognizing *other* forms of life, and understanding that we all have something deeply fundamental in common, even though much of this other life may be without use to us or actually dangerous, like snakes, tigers or certain bacteria.

Wilson wasn't the first person to have this idea. More than a century earlier, Henry David Thoreau, the American philosopher, nature poet and avid hiker, had written about how humans had the capacity to transcend the mind–nature divide. In one sense we're separate from nature because we can observe it, measure it and think about it, but at the same time we're a part of that very same nature. So as we're observ-ing it, we're also observing ourselves. If we think and feel in this way, it's easier to accept the idea that all living things are made up of some shared substance.

Through this approach, Thoreau found a hope that we might feel more deeply for nature and care more about it. The Swedish writer and thinker Sara Lidman revisited a similar idea. She coined a Swedish neologism, *samvettet*, akin to a new dimension of the existing word 'conscience' in English, which implies that this experience of collective belonging touches something of a moral centre within each of us. *Vett* is not just 'knowing', it is also a word for common sense, a word that signals what stands to reason. A kind of *con-science* to do with the experience of a reciprocal equality innate to all humans, not far away from Wilson's biophilia idea. It implies a kind of interdependence, for reasons and through forces that we cannot always rationally discern or appreciate.

Can snow be part of what our *greater conscience* includes?

Recognizing our connection with the natural world around us has long been a necessity. Modernity, industrialization and urbanization imposed a distance from nature that we've exploited without thinking. Now we've wisely started questioning that strategy. But it's still not enough. Our conscience asks us to examine the rationality of the behaviour we engage in. It's a form of knowing – *vett* – while the opposite is a kind of *vettlöshet*, a senselessness, where one doesn't fully understand what would be obvious to any normally empathetic human being. Thinking in these terms, it's not a stretch to feel sadness and loss when faced with melting snow. And that's how people feel. In Greenland, with its substantial ice sheet, which is mostly covered by snow, almost half the population say they're fearful of climate change. The extent of the snow cover has also decreased significantly in Greenland, and the ice sheet on the huge island is melting away faster than ever before.

In an article from October 2020, the *Guardian* suggests that populations living in the Arctic are experiencing 'solastalgia'. The term comes from Australian social psychologist Glenn Albrecht and is related to 'nostalgia'. The root of the latter part of the word, *algos*, is the classical Greek for pain. But while the antecedent *nostos* in nostalgia means 'homecoming', and nostalgia as a whole refers to a 'wistful, but pleasurable longing' for something lost, the antecedent in solastalgia comes from the Latin *solacium*, meaning 'comfort'. Solastalgia is also related to the English word 'desolation', a blend of devastation and despair. De-solation. Without comfort.

Albrecht put forth the concept of solastalgia in 2003. It's taken time, but solastalgia is becoming increasingly established among professionals in psychology, landscape conservation, memory research and people trying to understand the current state of our emotions. The concept seeks to capture the painful longing of nostalgia for a particular place, the way it looked before, while at the same time experiencing awareness that the object of this longing has irrevocably lost the very qualities we seek, due to climate change and environmental degradation. It is, in Albrecht's words, 'the loss of, or inability to evoke, consolation', the very kind of consolation that many people would seek out in familiar environments. It's a bit like the feeling that would afflict a person sentenced to be forever at sea, unable to return to harbour. This was once considered the worst punishment a person could receive, worse than serving the equivalent time in prison on land. No smells of home, nothing familiar in sight.

Those affected by solastalgia may already be refugees from homes to which they cannot return because they've been flooded, crushed in poisoned landscapes where animal and plant life have been destroyed or transformed, where certain birds no longer sing, where thick vegetation covers what were formerly open vistas, or where clear-cut forests with dried-out litterfall and furrows ploughed for replanting mean that the term 'forest' no longer matches the reality before them. Some places, like the Chernobyl nuclear-disaster site, are closed off and restricted. Others are nominally accessible, but in practice uninhabitable, or so unattractive that no one wants to or is able to live there any more: the atolls affected by nuclear testing in the Pacific, the salt desert of the Aral Sea. Where the rainforest has become savannah in the Amazon, the indigenous population has effectively been displaced without there having been any conscious decision to expel them. The destruction of one's home environment doesn't have to be fully complete for these feelings to arise. Like other forms of anxiety or concern, the idea that we're approaching this kind of state, and the powerlessness in the face of the forces taking us to it, is enough to induce the singular pain of solastalgia. We can think again of the 'nuclear winter'; it doesn't have to have transpired in order to throw people into anticipatory distress.

Part of one's home environment includes the characteristics and elements related to seasonality and climate. The original features of places, the things that have shaped the living beings residing there over time, are as much a part of the place as geological formations and the angle of sunlight during the different times of the year. What we loved no longer exists. And for many of us, what we loved is snow, and its absence has started to evoke in us a state of sadness. The objects of our solastalgia are fungible, in the same way that the world we've lived in, no matter where in the world that might be, has been changing.

It can, of course, be argued that the earth has been in a constant state of change throughout its more than 4.5 billion years of existence. But the concept of solastalgia isn't referring to the kind of relativism that stems from such reductive common sense. In the face of these vast time periods, a person might experience awe, vertigo, humility. But solastalgia is tied to the human timescale, which encompasses our lives and experiences as humans. It's a feeling, an experience, of pain. It is about something close by and intrusive, the pain we inflict on ourselves through our own impact on the world. Nuclear accidents, deforestation and species extinctions are not distant abstractions, nor sudden natural disasters. This is the work of human beings.

Solastalgia is a concept that belongs to the Anthropocene, to our own period, when the pace of change has accelerated such that natural variations pale in comparison. In 2023, the earth's average temperature rose from 1.1° Celsius above pre-industrial levels to more than 1.5 degrees above. That kind of an increase in a single year is unprecedented in the earth's modern history over the past million years. We're on thin ice.

'The earth is faster now,' announced Legraaghaq, a tribal elder of the Yupik people. It was his way of describing the retreat of the sea ice and the melting of the land's permafrost. Legraaghaq has spent his entire life in a borderline existence. The Russian–American border is nearby, as is the international dateline, which means that two days are always visible at any given moment.

But human life and the environment in all its dimensions are changing now more than ever. During her long-term research among local populations around the Bering Strait, both in Siberia and Alaska, historian Bathsheba Demuth has often encountered tales of alienation as everything accelerates. The limits of sustainability, too, are being exceeded. Legraaghaq's experiences are typical ones in the Anthropocene. The title of Demuth's book, *Floating Coast*, is apt precisely because what was once solid has become fluid: the snow's arrival, which is delayed; the snow's depth, which is ever-decreasing; and annual animal migrations, which are increasingly irregular.

German psychologist Hannah Comtesse studies the long-term effects of war, displacement and migration. She has also published studies showing the extent of the anxiety, fear and despair that people around the world are experiencing as a result of climate change. The people of Greenland feel not only fear, but also sorrow; between 5 and 7 per cent of the population also experience hopelessness or anger, in varying constellations. More than half (56 per cent) of rural Australians are also concerned about climate change. In the United States, the corresponding figure is closer to a third (29 per cent). Almost two thirds of Americans are 'very or somewhat concerned' that extreme events will occur in their own neighbourhood. Similar sentiments exist in Europe: a study conducted in France, Germany, the United Kingdom and Norway indicates that around 30 per cent of people in these countries are 'very or extremely concerned' about climate change, while feelings of fear, anger and guilt range between 8 and 42 per cent. New terms like 'ecological grief' have come to the fore. A study in the medical journal the *Lancet*'s report 'Countdown on health and climate change' referred to ecological grief and even 'ecological anxiety' as a 'healthy response' to climate change.

Words matter. Psychologists have emphasized that grief and other emotions related to anthropogenic climate change are easier to discuss when there are terms for them. Words, especially in the form of diagnoses, give sufferers confidence and can also help them better recognize and describe their symptoms. But the phenomena that sufferers are concerned about aren't always clear-cut. Just as a few tenths-of-a-degree increase in temperature per decade cannot be perceived by a person with their own senses, neither can a 7 or 8 per cent reduction in snow depth. The annual variations are too great. Memory is too fragile. You need measurements and data. In general these are perfectly reliable, yet they are constantly being challenged, often quite cynically. Donald Trump is one of the people who regularly disparage the scientific data. Systematic smear campaigns against those who put forward uncomfortable messages are numerous and well recognized, and have now been thoroughly studied. But doubt is also expressed closer to home, if more quietly, by sceptical or uncertain family members, neighbours, colleagues. Is climate change really *that* bad? Our situation isn't really so difficult, is it? And does it matter if the weather gets a little warmer?

It gets even more difficult if the negative feelings coming up aren't rooted in diagnostic language. This is why words are important, and so is a public dialogue about loss. If you suffer from the desolation of solastalgia because the place you love no longer exists, you need to do some form of grief work. In so doing, you can take solace in others around you in the same predicament. Panu Pihkala, an environmental theologian at the University of Helsinki, has spoken about how many Finns and, for that matter, other people in the Nordic region experience 'winter grief', known in Finnish as *talvisuru*. Grief support groups have formed that rediscover older ways of expressing and processing grief. These groups like to gather in places linked to the losses they're experiencing, expressing their feelings through ritual, as the practices of mourning have been done throughout the world for as long as we have recorded accounts of grief. In the UK, there are societies that come together to practise 'good grief'.

Women seem to find this kind of work easier than men, perhaps because caring and nurturing, things that are traditionally female-coded, are so critical for people affected by brutal change, and working in resource extraction is more likely to be male-coded. Electoral statistics in many countries show clear differences along these lines. Women prefer political parties that have caregiving, environmental concerns and the climate high up on their list of priorities. Many men nowadays vote for national conservative populist parties that hold views sceptical of climate change and that operate in an emotional register characterized by cynicism, contempt for weakness, and embracing of values that defend and legitimize the high emissions of greenhouse gases and environmental toxins.

In other words, grief arising from these feelings of loss is unevenly distributed across the population. This divides people. Those who act morally and emotionally against climate change may be in the grip of rage, their first impulse being to rail against those who continue to resist reform and emit pollution, but by extension also against the kind of social order that allows this work of destruction to continue. Glenn Albrecht has found a term for this too: *terrafurie* – literally, 'wrath of the earth'. Or, perhaps more accurately, anger on behalf of the earth, all the more intense because the earth itself cannot speak. You can sense this anger behind climate action such as road blockades or oil pipeline occupations organized by climate protesters in the United States.

On the other side of the coin, there's rage among some motorists and air travellers whose transit is impeded, even if only temporarily, by the protests. What characterizes these actions is that they actively expose climate-damaging behaviour. And given that this behaviour is founded on an entire techno-economic world order, ultimately predicated on cheap fossil fuels, many individuals are outraged when their active participation in this world order is criticized. Maybe the true impact of their own behaviour dawns on them in these situations, in ways that they rarely find very pleasant. Hence the rage and cynicism, helping to tamp down feelings of guilt and responsibility – but at the same time revealing those very emotions.

But it's also possible to seek understanding. Not all tractor-driving farmers – who also block roads in protest, but for completely different reasons – are climate-change deniers. It may be that not all of them even have a settled opinion on climate issues. But it's likely that many of them count themselves part of a group that has been farming the land for generations. Maybe they see themselves as having been stewards of it. The growing number of studies of rural life show that many have experienced deep-seated disappointment after witnessing generations in which cities have grown in power and resources, and continue to do so. And many of them aren't being listened to. Their welfare and life opportunities have diminished. Why aren't these losses counted? There are many of them: access to a local post office, healthcare centre and emergency services.

Panu Pihkala of the University of Helsinki has more to say. He, too, sees an ecological nostalgia of the kind Glenn Albrecht talks about. People who have no influence over their situation – which is a difficulty suffered by people missing the experience of snow – can fall into a state of melancholy, a passive longing for a past state that will never return. A sense of powerlessness that, in the more severe cases, can turn into despair. This is a state of mind not unlike that of a person mourning their dead child, or someone who's been abandoned by a lover.

'Snow anxiety', the Finnish *lumiahdistus*, happens when someone doesn't know if there will be snow in the winter. The word 'anxiety' is justified, says Pihkala, since 'problematic uncertainty carries with it anxiety'. The condition can affect both the young and the old. The lack of snow also increases anxiety by increasing winter darkness, with potential physical and mental consequences. *Talvihaikeus*, the Finnish word meaning 'snow relief', on the other hand, is a form of cheerful sadness that can be associated with winter – *talvi*; *haikeus* is a word that expresses both sadness and gratitude. Something is about to change, perhaps disappear, but what remains is still valuable. 'Winter joy' – *talven iloa* – and 'snow relief' denote the special happiness that comes with the arrival of snow, because it's not something that we take for granted any more. When it does come, it's even more welcome than usual, because there's an entirely plausible alternative in which it doesn't come at all. So a little dusting for a brief moment is better than gluey mud all winter long.

It's against the background of this emotional activism and the mourning related to the retreating ice and snow that we should understand the social-psychological effects of reduced snow. These are multi-dimensional, causing sorrow to both geographic areas and to individuals. Some people are more sensitive to this, others less so. Things are also developing all the time, and difficult to assess. The vast majority of people encounter climate action through the media, which portrays it in different ways, often depending on the media's position in an increasingly politicized public sphere.

This doesn't necessarily mean that people don't have feelings about it. As we have seen, much of the population in the northern hemisphere, where snow is the norm but declining, is concerned about climate change. Not so much about how it impacts the water supply, ecosystems and hydro-electric energy production, important as these things are; their concern is about something bigger. It's about the seasons. The light. Our senses. Something we can't fully define – except that this wasn't something we intended to create. It's at this point that we start thinking we should have left the white whale alone in the depths.

Cold Activism

*

SOMETHING DREW ME INTO THE SNOW TUNNELS. We dug them longer and deeper. Scooping out all that snow provided a strange sense of satisfaction. I didn't know much about this word then, but it was *work*. It was its own reward. I was stubborn. Once I'd begun, I could barely stop. It was quiet. And not as dark as you might think. Even though the daylight outside was weak, what light there was managed to make its way through the entire mountain of ploughed snow, illuminating the secret world we had created for ourselves inside it. I wore ski boots with yellow-and-black laces and a thin brown jacket that my mother had made. She always did her utmost. Most people had store-bought jackets. My jacket was actually a little too thin for the wintertime, but I never mentioned that to anyone. The snow froze in clumps on my mittens.

There inside the holy crypt of snow, I was safe from snowballs. But not from the bodies of the other kids, who, during their snowball fights, took up all the space in the tunnel, so that you couldn't get past them; or from the way that words lost their meaning when you were divided into two teams in a snowball fight. I didn't understand where the terror came from. It filled me, coming from an unknown place deep within myself. Filling my whole body as if it were dark water from a secret spring. All the way up to my head, where all memory ceases.

Maybe that's why, much later, I had so much anxiety when I read *The Ice Palace*. The novel by Norwegian author Tarjei Vesaas is about an eleven-year-old girl named Unn, who disappears. My anxiety surprised me, but it made me remember those tunnels we dug in that time before time. Unn disappears when the autumn chill has set in, but it hasn't yet snowed. It's only when the villagers realize she's gone that the snow starts to fall.

> There had been a cold, but marvellous skating season for a long time. Today it would end, today the snow would come.
>
> When they went out of doors after the next lesson the ice had already begun to whiten. [. . .] Extraordinary how quickly a thing can be destroyed. The ice was flat and white and dead.

They start searching for Unn. And soon they come to the ice palace that's formed by water freezing at the spot where the river cascades down out of the lake. Inside the ice palace is the same silence and light that I remember from the snow tunnels we dug. One day, when Unn's best friend Siss climbs into the ice palace, she catches a glimpse of Unn's face behind the wall of ice. Larger than it was in real life. Indistinct, without contour, at the same time unambiguous and unreal.

Unn is never found. It keeps snowing: 'The great lake was a silent expanse, no longer detonating, nonexistent.' Vesaas's book was published in 1963. The Swedish translation that I read came out in the literary promotional series *En bok för alla* ('A Book for Everyone') in 1982. On my bookplate I've written "83". Almost exactly when my own problem started to become clear to me. Almost exactly when I heard the ice in Rondane.

Activist events dealing with snow have to date been relatively few and peaceful. We have seen far more of such events to do with ice and glaciers, probably because glaciers remind us more of monuments in the classical sense. They are 'alive' in a way, in the same way that rivers are alive. Ice has plasticity and tenacity. Gravity moves it slowly from higher altitudes to lower ones. Glaciers are also constantly given new life through the new snow that replenishes them each winter. And they have a clear location, bounded topographically by mountain ridges and valleys, then by the warmer temperatures at lower altitudes. A glacier is a specific and defined object that has kind of an identity, often a history. Glaciers have local names. They've been documented and catalogued, visited by mountaineers and scientists.

Glaciers contain material that calves off and melts, revealing things like old trees, vestiges of human activity, stones and stratified objects. Even macabre items contribute to humanizing the ice, tying it to its place in history. In the Alps, there's thawing of both the old (5,000-year-old Ötzi) and the new (alpinists and slalom skiers from the past fifty years) bearing witness to human presence in the grand cycle of freezing and thawing. It's the same in the Himalayas, where unfortunate mountaineers on Mount Everest are occasionally extracted from the ice and where excrement litters the slopes. As snowfields and glaciers melt, elements of human cultures emerge, everything from ancient arrowheads in Norway to mummified remnants of sacrificed bodies at extremely high elevations in Argentina, and plutonium recently discarded in the Himalayas. There has been a scientific journal for glacier archaeology since 2014; such research is also ongoing in the Swedish mountains.

When Argentina wanted legal protection for its glaciers, the first thing it did was to launch a research and inventory programme, resulting in a national atlas of all the country's thousands of glaciers. A Canadian mining company wanted to blast mountain areas to access high-altitude deposits of gold. The Argentinian legislature required the company to supply the name, size, depth and climatic conditions of each glacier at risk of being affected, so the sites could be protected and preserved. The legislation in 2015 was prescient and unique. But it still wasn't enough, because the force that is ultimately the threat to Argentina's snow and glaciers is atmospheric global warming. The local – the legal – protection is symbolic, though it's important even in that limited capacity.

Tentative explorations along the same lines are taking place in many parts of the world. The World Bank funded a project in the Andes in which mountain ridges were painted white to raise the albedo, reflect more sunlight and preserve the glaciers and snow that lie between them. It's been suggested that Kilimanjaro's melting glacial summit be covered with light-reflecting blankets, to the same effect. In 2004, Greenlandic conservationists succeeded in obtaining World Heritage status for the ice sheet and fjord at Ilulissat on the west coast of Greenland, where meltwater and calving ice are occurring at ever-faster rates. Others have fantasized about using robots to build dykes on the seabed or installing giant 'underwater curtains' to stop warm seawater from reaching the glacier front or the shelf ice, pumping cold water into glaciers to rebuild them, spraying artificial snow over Antarctica, or sending submarines to the Arctic to spearhead the desalinization of seawater and create artificial icebergs. Alongside deep, solemn mourning, one can see this as a technological escalation using hard-boiled engineering methods to protect the retreating cryosphere. The American historian Mark Carey has called these attempts 'defence campaigns' and has examined how glaciers, along with the planet's ice and snow, have become an endangered species of global proportions. So, do these campaigns work?

There are ancient techniques for preserving domestic quantities of ice and snow, including insulating it inside root cellars, or under sawdust or other coverings. These methods were dominant until the advent of modern refrigeration technology in the 1920s, which made natural ice and snow largely redundant as a cooling material within a matter of decades. This assumed that large amounts of energy could be supplied to maintain the 'cold chain', as it was called, with mobile refrigerated trucks, freezers and refrigerated aisles in stores used around the world. The focus on natural ice soon shifted to the technical development of materials that could preserve ice in its natural state, especially glaciers. In the 1960s, new technologies were introduced in the form of thin light- and heat-reflecting substances, along with insulating foam. These kinds of materials could also easily be used to preserve heat, and became important in the treatment of accident-related injuries.

Around the turn of the millennium, when climate change started to be more evident in major wintry tourist areas, that demand started to come from ski resorts. At the Stubai Glacier near Innsbruck, trials were conducted using blankets made of hemp and wool, and for the past twenty years engineers have been using blankets made of polyester or various sugars, and other geotextiles. A certain desperation has become noticeable in the Alpine countries' winter sports industry. This is a multi-billion-euro industry that has long attracted upwards of 100 million visitors annually (although the pandemic reduced those figures somewhat). But according to a study published in 2019, hundreds of ski resorts have now closed, primarily due to lack of snow, as a result of climate change.

One of the most famous of these was the Hotel Belvédère near the great Rhône Glacier in Switzerland, built in 1882 by Joseph Seiler. A tradition of visiting glaciers and seeing snow and ice had emerged from the 'grand tour' trend in tourism. Young British, German and, to a lesser extent, Scandinavian travellers, mainly from upper-class backgrounds, would undertake long, often multi-year, educational trips to see the art and architecture of the Middle Ages and the Renaissance, especially in the Mediterranean countries – Italy being the prime destination. These trips also included detours to the Alps. The Belvédère was sited directly alongside the Rhône Glacier, so that the rooms for the ninety guests were right beside the main attraction. Seiler also had a tunnel dug into the glacier so that visitors could get inside the ice itself to experience its shimmering cold spectacle with all their senses.

The period known as the belle époque was the heyday of luxury hotels. In the Swiss canton of Valais, they multiplied within a few decades, numbering in the hundreds by 1914. The Rhône Glacier was further immortalized by Sean

Connery, who drove his Aston Martin through the hairpin curves around the Belvédère in the 1964 James Bond film *Goldfinger*. These days, the glacier's long tongue has melted away, its lustre has faded, and the hotel has been closed since 2018. Now it stands there like something out of a Wes Anderson film – which it very nearly is. The hotel, and the glacier, are on the cover of Wally Koval's 2020 book *Accidentally Wes Anderson*, already monuments to time and snow gone by.

Many people find it hard to take losses seriously, or to recognize that there even *are* losses, caused by our rampant, heedless economies. The world's attention has instead turned to the attempts to save glaciers from obliteration by covering them. Engineers, rather than tourists and skiers, have become the focus. What they can't save with snow and ice protection, they replace using snow cannons and chemical methods that compensate for the lack of natural snow. There are similar solutions in the other areas of the world where glaciers need to be saved.

In order to ensure, at least to some degree, the supply of fresh water to the major rivers of northern India and Bangladesh, artificial glaciers have been created in the southern Himalayas. The engineer Chewang Norphel pioneered the construction of artificial glaciers in the form of deep caves in the ground, in which water diverted from higher sources is then frozen. A saviour with a local solution to a global problem, Norphel soon became globally renowned as India's 'Ice Man'. Another approach is to build upon the local *stupa* tradition. *Stupa* are columns of ice, several metres high, that have long existed naturally. They are artificially created by directing a jet of water upwards during the cold season. This is done on a small scale and doesn't take much effort. Both of these ice constructions make their contribution to the fresh water supply when they start to thaw from April onwards and during the summer, leading to some improvements, though on a scale that is still completely inadequate.

A 2019 international study found that even if the Paris Agreement target, of a maximum 1.5 degrees of warming above pre-industrial levels, were to be met – which seems unlikely – one third of the ice and snow-mass in the Himalayas and Hindu Kush would melt. Freshwater supplies would be severely reduced for those 1.5 billion people, and almost twice as many would be at risk of crop failure in countries such as Afghanistan, Bangladesh, Bhutan, China, India, Myanmar, Nepal and Pakistan. This is a global issue with an ominous trend. In about half of the regions where people depend on meltwater from snow and ice, water availability has declined, with several regions already having passed 'peak water' – that is, a point at which water resources become constrained. In addition, torrential rainfall is likely to increase, while the quality of the meltwater itself is deteriorating as toxic substances, such as mercury, plus radioactive substances and PFAS, are deposited on the snow.

For people in the Himalayan region, the snow crisis is nothing new. The year 2024 was another in which the snowfall amounts were far below what's expected. As usual, language bears witness to the significance of snow and glaciers. In Sanskrit, Everest is called 'the Abode of Snow', and in Tibetan, 'the Mother of the Universe' – *Chomolung-ma*. Throughout the region, people have united across national borders for a campaign called SOS – Save Our Snow.

The outlook is bleak. Mark Carey and his team have noticed a pattern in the way climate mitigation projects work:

the rhetoric overshadows the results. Covering a few per cent of the Rhône Glacier may have reduced melting by one or two thousandths. The artificial ice structures of the Himalayas have produced slightly better effects, but they are still marginal. Despite being essentially useless, these projects attract a lot of attention, especially from the media. They focus on technical interventions and on the engineers, who are portrayed as innovative and heroic, without actually having solved any problems. At the same time, they draw attention *away* from local populations and the huge impact to jobs and the workforce that climate change will have, regardless of all the flashy rescue efforts. These efforts are mere plasters on injuries that are not just glancing, but gaping wounds. Political solutions to the climate crisis, which are the only thing that can really save the ice, get completely overshadowed in the media drama around local ice conservation.

Art projects have also been used to raise awareness and draw attention to the sad state of cold regions. These art efforts, now worldwide in nature, have sometimes been produced in tandem with the engineering projects' orientation towards snow and ice as elements, though usually with a different narrative framework. Icelandic–Danish artist Olafur Eliasson, working in collaboration with Danish geologist Minik Rosing, had twelve floating ice blocks removed from the Nuuk Fjord in West Greenland and shipped to Paris at the time of the COP21 meeting in December 2015, where they were laid out in a circle on the Place du Panthéon, in front of the monumental so-called 'palace of heroes', so that each of the ice blocks corresponded to the hours of a clock. The ice melted as time passed. The work was called *Ice Watch Paris*, and its clock formation shows we are merely observing the passage of time, unable to intervene; that the tragedy of change happens continuously and relentlessly, while people at the COP and other meetings mostly talk, and the futility of various more or less short-term projects becomes evident. Eliasson hoped that his work could bridge the gap between scientists with their facts, on the one hand, and politicians, world leaders and other people with their emotions, on the other. Author and activist Rebecca Solnit declared it to be 'A beautiful, disturbing, dying monument to where we are right now.' Humans themselves are melting the ice, but are unable to make the changes that prevent it from melting.

In the summer of 2015, on the summit of the highest Swedish mountain, Kebnekaise, Stockholm-based artist duo Bigert & Bergström laid out a 500m² reflective canvas as part of their project titled *The Freeze*. The yellow canvas was clearly visible from afar, particularly from the air. With direct reference to the Rhône Glacier, the artists wanted to simulate an ice-and-snow rescue attempt. Preserving the cold. Literally freezing the height of Sweden's highest point, the glacier on the mountain's southern peak. The canvas would reflect light back into the atmosphere to stop the melting of the glacier.

Unlike the glacier projects in the Alps and Himalayas, the artists' point was the futility of their gesture. The forces unleashed by people through their daily actions – though, as I said, on quite different scales – far outweigh any attempt to preserve any individual entity, such as a glacier. Bigert & Bergström's installation can be read as a protest against the endless and pointless 'war on weather'. Our mission should be to heal the earth: to lay down our weapons and turn off the taps on our fossil fuels. Beyond their fascination with technology, and the often strikingly beautiful, seductive and deceptive scientific imagery that Bigert & Bergström have used through decades of climate art since the early 1990s, they find nothing but emptiness full of prejudice – none of this effects real change or makes a difference. Which proves that there's nothing there. If anything, just more meaninglessness.

Bigert & Bergström's work on the 'Earth system' – the planet's functional systems on a larger scale – are subtle but clear signals about choices that societies now have to make. Should we keep going as we've been doing? Or should we try to break out of this vicious circle? The bandage – if you want to call it that – on the glacier also has ironic overtones. The scale is quite deliberately pathetically insufficient. Adding to that irony is that the following year, in the summer of 2016, the summit of the glacier had grown by thirty centimetres. Bigert & Bergström point this out in an article on the project website, tongue in cheek: 'Can art save Kebnekaise?' Of course it can't. In subsequent years, the peak's glacier has shrunk more than ever.

The message is now the exact opposite. The ongoing war against the earth is being waged on the largest possible scale, and is part of what we would term daily life: the routines of billions of people, the reflexive behaviour of companies and financial markets. Things that would be considered criminal if such damage were inflicted directly to other people, but which is condoned because the damage is aimless and unintentional. The rescue operations are paltry and futile, and are carried out on the lowest possible scale: on each individual glacier. But the impact of art is of a different kind. Art changes the way we think and feel. It moves us, overcomes barriers, often using pathways we don't fully understand. Things that help to shape the world anew.

But art can't do it alone. It needs support. How genuine are people's reactions? Does it matter if the ice disappears? Does it matter if drought spreads? Does it matter if storms get worse and more frequent? We're the only ones who can answer these questions, using as a foundation our values, based on our shared experiences and memories. All stemming from our fear that there's something precious that might be lost – the ice and snow that have always been present, the seasons that we learned to exist in. Whether we love or hate them, we've never been indifferent to them. Putting bandages on glaciers demonstrates the act of caring. We, too, are nature. And we're a part of nature that feels.

The Kebnekaise art project reinforces the monumental nature of the glacier. The southern peak had been the highest point in Sweden; now it's irretrievably lower than the northern peak. This is typical of a form of art activism that emphasizes the symbolic importance of a specific geographical point. The same monumentality has been useful in the memorial ceremonies in Oregon and Iceland. The glaciers that have disappeared are made present by their absence. I think this explains the attention that the annual measurements of Kebnekaise's southern peak receives in the Swedish and international media. You need a place, what the American sociologist Thomas F. Gieryn has called a 'truth spot', which can sensually and symbolically show the truth of a complicated context. A place that bears witness to that which is irrefutable. The glittering-jewel glaciers and the pristine white snow serve this purpose very well.

To understand presence and absence, we need to be able to experience ourselves as embedded in time. In change. The place where the loss is taking place is unambiguous; the terrain still bears traces of the ice. We would do well to see through those eyes that first saw these things for us. Such as the way the Norwegian geologist and meteorologist Jens Esmark experienced this in August of 1823. In a hurry to get back to the University of Kristiania to teach his autumn classes, Esmark takes a short cut across a chain of mountains, going as far as the northern tip of Jostedalsbreen. In a mountain pass he finds his way blocked by the ice of the glacier and has to look for another, accessible route. This leads him quite by chance to make a discovery that will change the world for ever. As he searches for a new path, he comes across residual organic material that has emerged from the melted ice, lying on the ground like ruins. He sees abrasions on the rock, colourations, boulders that the ice has carried along with it on its journey. He sees this material lying there, incomprehensibly discarded, with a completely different structure to the bedrock on which it happened to land. At the time, other people have witnessed the same phenomenon in many other places, but as far as I know, Esmark is the first to imagine that there was once an ice age, the effects of which explain the landscape he sees before him. This is Esmark's epiphany.

It's the same discovery that Louis Agassiz made in Switzerland scarcely a decade later, and a similar observation is made by the polyglot poet Johann Wolfgang von Goethe, among many others. The important thing isn't so much who was first, but to realize the importance of reinterpreting the familiar. Esmark's viewpoint was primed for change. He had measured the variation in temperature, the extent of snow. Ever since the moment he saw the landscape with new eyes, glaciers have become measurements of climate and, indirectly, of time. We now understand so much more regarding the presence and absence of snow and ice.

With snow, it is difficult to locate a specific location to focus grief or pain, or a specific time period to be recalled or revered. Snow isn't a place. It has a range, but is rarely a definable object. On the other hand, there wouldn't be any glacier ice without snow. It's in snow that ice begins. With ice also comes snow. Snow covers the ice for much of the year, from winter's downy, soft cover to the slush of late summer, in which snow and ice mix together with meltwater. And while a glacier may be fairly free of snow towards the end of summer, fresh new snow is added during what is usually the long winter season. The ice, despite its vulnerability, is the stable material. Snow is something ephemeral. Even in the winters of old, it came and went, and the thickness of the snowpack varied. It is true that the depth of snow has been systematically decreasing since around the middle of the twentieth century, but there are large variations within years, between years, and over time. Some of the greatest snow depths ever measured around the world have been recorded relatively recently; in Sweden, on the other hand, the greatest snow depths came in the early twentieth century. Nonetheless, the overall downward trend of recent decades is clear.

There's no obvious place for either the celebration or the mourning of snow, nor any moment to be particularly remembered or celebrated. No zero-hour can be established for snow as it can for a glacier. Glaciers have a continuity, an origin story, and a cycle of life, with highs and lows in their range and robustness. It's possible to write a biography for a glacier. But hardly that of a snowdrift – or a Scottish snow-patch. If snow lies around for a long time it stops being snow, gradually turning into ice, which may eventually form a new glacier. Even record-breaking snow cover almost always melts away during the following summer season. These days, with ongoing climate change, there are no new glaciers being formed. The only thing happening is that the existing ones are dying.

Snow doesn't die. It dwindles. Its presence becomes more and more irregular. It flickers and flutters, the way things do when they're petering out. But it rarely stops for ever, and if it's absent one year, it might come back another. Its features are there, though they're increasingly unreliable. Snow droughts are growing in scale and frequency, yet there's still no acknowledged memorial place. Even if there were such a place, it takes only a few large snowfalls during a cold winter to quickly cancel the drought. A glacier, on the other hand, is more like a candle: once it's out, it's out. It takes many years to be rekindled.

It's not really possible to say that snow is gone for ever in any individual place, or even that future winters will now always be winters in which there's little snow. The decline is a question of averages over time. There's still the occasional especially snowy year. Even in places where snow is rare, it still sometimes appears. There can be snowflakes in the air in Rome in late spring, as I have personally experienced. In February 2022, heavy snowfall in Texas paralysed the city of Austin. The USA's National Snow and Ice Data Center in Boulder, Colorado, contends that most of the North American continent has seen snowflakes in the air at some point, if only at great intervals, as in Florida. Meanwhile, huge snowstorms and the odd winter with an abundance of snow give climate sceptics and other disbelievers a few drops of water with which to whet their homemade millstones.

What's happening on a grand scale is happening quietly. But people want a smoking gun to point to – a symbol, a place to gather around. The fate of ski resorts may be the answer. Bolivia's only ski resort, on the Chacaltaya Glacier, closed in 2009 when the glacier finally disappeared after receding for twenty years. Even in Whistler, the ski paradise of British Columbia, more and more of the lower-level precipitation falls in the form of rain, pushing skiing areas ever higher up the mountains. In more and more places, the day has already come when the lift has closed for good, when the hotel has to close down. Maybe that's because their public has given up on them – but that, in turn, is because the snow has been scarce for so long, and to such an extent, that the area simply isn't attractive any more. If you can't rely on the snow, the resort is also going to die. And if you can't rely on the snow, you can't really be sure about the land, either. The permafrost is thawing out and the 'eternal' high-altitude frost is dissolving. In the winter of 2024, the cable car in the ancient French ski resort of Chamonix was dismantled. It was no longer safe because climate change had weakened the ground the lift had been anchored to. Meanwhile, a new cable car is being installed higher up the valley to meet demand from tourists wanting to visit the French Alps' largest glacier, Mer de Glace – the 'Sea of Ice'. Half a million people come every year – to see the glacier melting.

Many other cable cars in the Savoie region have gone the way of the one in Chamonix. And in Siberia, the tundra is sinking as the permafrost melts, exposing mammoth carcasses and their bones, blood and DNA. This, despite being the coldest area on earth outside Antarctica. But that's not the full extent of it: the tundra has been freezing and melting, freezing and melting, over and over again for millions of years. British geographer Charlotte Wrigley has called it living on an earth of 'discontinuities', but now the discontinuities are increasing in frequency. Melting is accelerating. There's nothing permanent left. The elements we have come to rely on are dissolving before our very eyes. At the same time, in the new archives of nature that are opening up, literally under our feet, we can find detailed data on the times and speeds of climate events that happened in the earth's past.

Step by step, the cryosphere (we'll come back to this concept) is retreating, and everything we know suggests a continuous and likely accelerating trend of diminishing cold. Northern winter months in recent years are the warmest ever recorded. At lower levels of terrain, the snow itself is disappearing. Higher up, the infrastructure is loosening from its moorings. Everywhere, anxiety and uncertainty are growing – probably including some who will be travelling on the remaining cable cars, too.

As for funerals and memorial ceremonies, such as those for the Ok, Pizol and Clark glaciers, we haven't yet seen their like for the disappearing snow. Maybe it's not that easy to organize one. It will be a slow death, probably an endless one. What the snow needs is a story. We have a beginning, some 2.5 billion years ago. And we've had a long and relatively happy human coexistence with the snow for much of the Holocene. But since the second half of the twentieth century, we've been heading into a snow crisis, deepening at the speed of an avalanche. What will the next chapter hold?

It's called a *lappkast*. I learn it first from my father, but more formally from my PE teacher. Everyone needs it to be able to ski.

We learn every aspect of skiing: uneven surfaces, ski jumping (if only at a very low height), snowploughing, *stämsväng* and telemark – including the art of creating tracks in the deep, new-fallen snow. Three students are positioned slightly outward at the front, in a fan-shaped arrangement, followed by three more, then three more – we're shaped in a zig-zag column, like the pattern on the back of a snake, all sharing the task of tamping the snow down firm and hard. At the end is the row of those who are creating the track itself.

Everyone is needed.

There aren't any snowmobiles yet. All right, there are – an orange one from the district of Ockelbo, used by electrical engineers when they're repairing the telegraph wires. I've seen it with my own eyes. But it can't replace the moral education that forming the ski track gives us.

We're eleven to twelve years old and up on the mountain, far from school and society. But this is still a lesson. This is where I learn that together we can make tracks for skiing. Tracks that will be used by other people in the civil society of snow. And we learn the *lappkast* – because there are times when the right thing to do is to turn around.

We're spread out across the snow. We each lift our left ski. Turn it in the air so that it faces backwards. Put it down on the snow. A small collective learning together how to turn around.

The Crystal Seekers

*

ON BREAKS AND DURING FREE PERIODS, some of us go skiing. I'm learning that I need to be able to read the snow to choose the right ski wax. The snow changes, day by day, hour by hour.

I get a little bigger. Shovel the snow in front of our house. A one-storey place by the River Ångermanälven. The best snow is the kind that's light enough for me to shovel it. To me, this snow is elegant. The overlapping layers growing ever bigger. Both my mother and father work at the school I attend. I get home first. I get to stay outside, and I don't want to do anything else, because there's snow.

To understand snow is to understand something 'central to our climate, the Earth and life. Understanding its behaviour is crucial to predicting the future of our planet.' This is what was stated in the journal *Nature* in the article 'Ten things we need to know about ice and snow' by Thorsten Bartels-Rausch of the Paul Scherrer Institute in Switzerland. Bartels-Rausch is one of the contemporary global scientists, many of whom are in Switzerland, trying to understand the most minute components of snow and ice at molecular and atomic levels. He and his colleagues are convinced that their work is urgent. They study snow and ice at levels of detail never before possible, and what they're finding is consistently increasing diversity and variety.

As I absorb the research from the Scherrer Institute, I start to think that the day may be coming when we have to stop talking about snow and ice as singular things. We might need to talk about 'snows' and 'ices' instead – that is, different kinds of ice and snow. And not simply because they have different ages, temperatures, consistencies, and other kinds of characteristics. But also because the group constituting 'ices' are also different at a deeper crystalline level. It's the same with snow: crystals can look very different, and are now being categorized into a dozen main types. 'We're only just getting started,' says one article, referring to our understanding of how these complex crystals behave. One-off experiments aren't enough; 'a long-term commitment' is needed.

Snow freezes and melts near 0° Celsius. Billions of people around the world live in or near environments with this sort of temperature. The snow might be found in distant mountainous regions, but large urban centres and vast agricultural areas can be dependent on its meltwater. Snow and ice, writes Bartels-Rausch, are 'polycrystalline materials . . . of particular interest because of their geophysical importance'. With climate change, such 'zero breakthroughs' – through the freezing point – will occur more often, and the areas in which they happen will expand further north and further upwards, to higher altitudes in mountainous regions. Freezing and melting will increasingly overlap, and the stable seasonal changes that have been the historic norm will grow rarer. This, in turn, means that snow coverage and depth, and also the age and composition of snow and ice, will become indicators of change and will affect how we, very literally, view our position. Time moves into geography, and vice versa: geography moves into time. We take on new chrono-geographical narratives about ourselves and our relationship to the places in which we live – and to other places, too.

So, snow and ice molecules are transformed to a grander scale, acting as a prism for the speed of time. Through the crystals, we can read the age of the frozen parts of the earth, the glaciers and snow plains, and the permafrost. And by comparing them with the snow and ice landscapes that exist today, we can study how they have changed and, in a sense, peer into the future. The snow forms a time telescope that can focus both backwards and forwards.

Maybe this doesn't sound that incredible any more. Isn't that what we expect from contemporary science? To be able to use some substance, almost any one, and extract from it information about the passage of time and the changes that have taken place on Earth? This was already the case with fossils, petrified trees (trees that have become minerals), lake beds, and other natural repositories. New insights into nature, environment and climate have increased the number of eras we define, most recently through research on the Anthropocene. This research has identified many indicators of human impact leaving lasting marks on the earth, in the oceans and in the atmosphere, and with an accelerating impact on the environment and climate. The ice sheets and seabeds have become repositories of the earth's recent history, just as the rocks and minerals of the earth's crust have become repositories of its ancient history.

So, isn't it only natural that it is in Switzerland that the role of snow in this drama is being studied? Isn't it precisely in this most Alpine of countries that we have reason to keep an eye on the snow, given all the winter sports resorts and the deadly avalanches that constantly and increasingly thunder through the valleys? This is true. We know quite a bit and are learning more all the time. But nonetheless, it's a little strange, isn't it – that snow, this airy, light, impermanent material that comes and goes with the weather and seasons, can carry this much information? How can we even study snow? As soon as I touch a snowflake, I disturb its fine crystals. When I carry it with me indoors, it immediately melts. No sooner have I hung up my coat and scarf in the hallway at home than there's a wet puddle collecting around the boots I've just stepped out of.

How does one come to know things about snow? How was it done in the past? How long have people been studying snow as an element and what have they been wanting to know? Much has been known for a long time, such as the fact that snow is found at high altitudes and that it melts with slight warming. So it was important that the melting point of snow, which is the freezing point of water, was set at 0° and the boiling point at 100°. It's been known for some time that thawed snow becomes more brittle when it refreezes. But many of the things that are most important to know about snow are those we're only just beginning to realize. Most of all, that it's a major common concern. Something that belongs within the political sphere, too.

A remarkable amount of the new information concerning crystals has been difficult to understand. The word crystal itself comes from classical Greek. The root of the word comes from *kryos*, which means 'cold'. *Kry'stallos*, in turn, means both 'ice' and 'rock crystal'. Maybe you don't think of 'crystals' when you see snow. But at its core, snow is crystalline, and it remains crystalline when it's transformed into ice. Crystal is matter that's organized in a particular way. And so is snow. But snow isn't crystal all the way through. The crystal is concentrated in many small areas called grains. Each grain is a crystal with an ordered structure. If you have a number of these kinds of grains, you get a multicrystalline substance. This is what Bartels-Rausch meant when he said that snow and ice are 'polycrystalline', in the same way that salt and diamonds are.

Another of the world's leading authorities on snow crystals is Kenneth G. Libbrecht, an astrophysicist at the California Institute of Technology in Pasadena. Libbrecht became interested in snow crystals in the late 1980s. He realized that people had been reflecting on these kinds of crystals for millennia; their hexagonal shape had been recognized in China at least since the philosopher Han Yin wrote in 135 BCE, 'The flowers of plants and trees are generally five-pointed, but the flowers of snow, called ying, are always six-pointed.'

Libbrecht recalls that as he delved deeper into snow-crystal research, he was struck by how different it was from other research he'd been involved in before. In areas such as astrophysics and material sciences, which he had been involved with for a long time, he was used to a research field having a clear organization, comprising journals, societies and conferences, and a number of unsolved problems that the researchers, who knew each other well, worked to solve. This wasn't the case with snow research, he says. I think Libbrecht exaggerates a little; snow research in the broad sense probably had its own epistemic community – that is, a common body of knowledge and its practitioners – and even the odd journal in which to publish. But whatever the case, that's the way he saw it, and compared to many established disciplines, snow-crystal researchers made up a very small community.

Libbrecht's own research covers the molecular components of a snow crystal, or 'snow star', as we usually refer to it in Swedish; how they're formed and how they organize themselves into the snow crystal's singular shape. At first, he felt isolated in this area, as though he were groping in the dark. Scientists had been working with this issue for a long time, but in the half-century since the late 1930s, up until the early 1990s, not much had been achieved, Libbrecht thought. And there's something to that: neither the battles of World War Two nor the Cold War would be won with greater knowledge of snowflakes; knowledge was sought in other cold realms instead – the stability of the tundra for landing aeroplanes, for example, or how the Greenland Ice Sheet would react if you dug into it and built a small 'city' inside, the way the Americans had with the Camp Century base.

By the time Libbrecht got excited about snow crystals, the Cold War had just ended and research was able to proceed in a new direction. Unsurprisingly, much remained to be done. His predecessors had categorized different types of snow crystals successfully. But the snow crystal itself, the invisible building block of every snowflake, still remained a mystery. 'I was immediately struck by how little was actually understood about how snowflakes were formed.' There was a great deal of knowledge about many other properties of snow. But nobody could explain exactly why a snow crystal looks the way it does, or why it looks different depending on the air, at different temperatures, or with different atmospheric conditions in different places. Or how it initially forms up there in the clouds.

There was one more thing that bewildered Libbrecht. When he lectured on snow crystals, he often received questions about who was funding his research, and what benefit there could be in researching snow. He'd never been asked these kinds of questions when he lectured on astrophysics. He realized that behind the questions there lay mistrust. What was the point of snow, really? As if something as beautiful as snowflakes was inherently suspicious. Fortunately, he was soon able to publish some popular books with evocative pictures of snow crystals. Some of these books he published in a partnership with the photographer Patricia Rasmussen, including *The Snowflake: Winter's Secret Beauty*, in 2003; beauty was still important – at least *that* hadn't changed. Some of these books were intended for children. These sold quite well and generated income that Libbrecht could use for his snow research. In doing so, he now had an effective response to the suspicious accusations he was sometimes subjected to: 'I'm not using one dollar of taxpayer money to research snow.'

Libbrecht's experience seems to touch on something quintessential about snow. Many people love it. Those of you reading this book may belong in that group. Those of us who feel this way say that snow brings out the best in people. I think that's because snow is an act of grace. After all, you could say that snow is really just water which appears as snow by mere happenstance – it happens to be both cold and full of fine-grained moisture up there in the clouds. Then it melts and disappears or turns to ice. What lingers is the memory of that magical state.

It reminds us of salvation. Many people know what salvation is because they think they've experienced it. But they don't know how to demonstrate what it is that they've experienced. The experience is difficult to make fully fathomable to others, despite it being often powerful and profound. This exclusivity of this individual experience is part of its unique significance. And the last thing a person who's been saved wants is for their tax dollars to go to someone explaining, once and for all, at the molecular level, what comprises that salvation. Salvation is reminiscent of the snow experience because both seem to involve a promise. Of what? I think it's about transformation. The snow carries with it a utopian message, a wordless promise of another possible state, both within ourselves and for the imperfect world we live in. If that sounds like magic, maybe it's because of that magical quality. Maybe we're not meant to understand everything about snow, any more than we can understand everything about salvation. And yet there have been many attempts. It's a paradoxical story.

Long before the philosopher Han Yin, the classical Greeks were already paying attention to snow crystals. Aristotle, who lived in Athens 2,400 years ago, is reported to have said that anyone who wants to appreciate the beauty of a snowflake must endure the cold. Aristotle was a scientific omnivore and enjoyed doing research outdoors – looking, feeling, smelling and tasting everything. His interests included snow, and he wrote about it in his work *Meteorologica* (parts 10–12) – 'Meteorology', which directly translates as 'that which is in the air'. Aristotle considered hoarfrost, which is rare, to be frozen dew, which there isn't much of either. Snow, on the other hand, was sometimes abundant in 'cold places', he wrote, and was formed by large clouds that contain a lot of water and produce snow when they freeze. He never explicitly says that the snow crystal is a hexagon, but he thinks it is beautiful. He is most concerned with hail, which in most of Greece fell much more often than snow. 'In general, hailstorms are more frequent in warmer places, snow in colder ones.' He was also interested in the mysterious phenomenon known as 'pink snow'. Sometimes the snow was coloured. Why was this? How did it happen? Aristotle realized that it must come from the air. But from where?

Another classical writer, Pliny the Elder, writes in the middle of his first-century book *Naturalis historia* about 'hail, snow, hoarfrost, fog, dew; the form of clouds'. He studied Aristotle closely, as many others were to do through antiquity and the entire thousand-year-long Middle Ages. There isn't a lot of writing about snow surviving from that era, however. The explanation for this modest treatment of the subject is that snow was relatively rare in the areas where learned texts were produced, and of little importance to most people of the time. Snow isn't a major theme in ancient texts, and where it does occur, it's mostly in the distance. In Euripides' drama *The Bacchae*, the mountain Kithairon, 'where the shining white snowflakes never melt', is usually seen only from afar. Similarly, the Greek geographer Strabo refers to the 'snow-covered' high mountains of Parnassus and Helicon. Strabo also knows about avalanches, that snow forms tiers and that it can fall from high altitudes, sweeping away almost everything in its path – but otherwise he doesn't have too much to say about snow. The historian

Herodotus describes the northern Scythians as speaking warmly of snow: 'Now snow when it falls looks like feathers, as everyone is aware who has seen it come down close to him.' Snow appears only in passing in Homer's *The Iliad* and *The Odyssey*. Plato mentions snow in his dialogues mostly in pedagogical similes. The philosopher Epictetus, in one of his *Discourses* lectures, makes an amusing comment about Diogenes, the learned ascetic who spent his life in a barrel. He is said to have appreciated the opportunity to train his own agility and physical strength by climbing on sculptures after a snowfall, as the slippery snow and the freezing cold made the task particularly challenging.

It is rare, too, that it actively snows in the books of antiquity, although Flavius Josephus mentions, in his work *The Jewish War* (c.75 CE), that Herod conquered the city of Sepphoris in Galilee 'during a very heavy snow-storm'. In *Anabasis* (c.370 BCE), Xenophon describes in vivid detail how Cyrus's army marches through deep snow and in fierce winds during the campaign through Persia. Xenophon, who was himself one of the ten thousand soldiers, describes how the army confronts the harsh winter in the Armenian mountains. The men become snow-blind, their toes freeze, 'rot' and fall off their poorly clad feet. The snow cover is so thick, in fact, that the fires they light to keep warm melt craters deep into the ground beneath them.

In general, snow only comes up in any detail in ancient literature in texts about war. It was there in the field that snow became a matter of life and death – as it was, for example, for Hannibal crossing the Alps with his army and thirty-seven war elephants, immortalized in William Turner's painting in which the approaching storm is just about to obscure the diminishing sun, as the soldiers crouch like Alpine azaleas upon the ground. The argument over which route Hannibal took is still in progress, but snow causing soldiers and animals alike to lose their footing, and plummet off the high passes down towards Italy, is depicted by both Polybius and Livy.

That aside, the word snow is usually used either to compare things, to emphasize whiteness – of a cloud, someone's hair, a woman's breasts – or in a figurative expression invoking innocence. Lucretius writes in *De rerum natura* (c.60 BCE) how nature is celebrated with 'the petals of roses falling like snow'; he sees beauty in the movement of snow. In the real world, however, snow is mostly negatively charged. In written works of the era, it is often associated with bad weather, unpleasant places and undesirable climates best avoided – probably because, for ancient peoples, snow was found in mountainous, barren terrain, or where it was very cold. In *On Airs, Waters and Places* (c.400 BCE), Hippocrates echoes other scholars in setting forth the idea that the most advantageous aspects of human beings were nurtured in moderately warm and dry climates, such as was the case in Athens. Hippocrates himself (along with his many co-authors in this collective work of somewhat obscure origins) has little use for snow. Melted snow or ice was unfit to drink, he held, and cold itself was not appropriate for humans, although he acknowledged that the meltwater from the mountains did contribute to fruitful harvests.

While interest in snow was limited throughout antiquity, it didn't grow much greater in the Middle Ages, despite the gradual shift northward of the centre of gravity for European intellectual life. Biblical motifs and religious allegories dominated painting, and church buildings were decorated with sculptures of prophets, popes and saints. Snow didn't figure prominently in any of the texts of the time, nor had it done so in the books of the Bible, where it's glimpsed as a term signalling something fleeting and rare. Visions of Paradise, which had become a common genre in the Middle Ages, never mention snow by name. Partly, of course, because biblical Paradise was a place of eternal fertility, and partly because snow was still – and was perhaps even more so than in antiquity – associated with plagues and hardship.

One exception is the whip-smart thirteenth-century Dominican monk Albertus Magnus, who had roots in Germany and is widely recognized as one of the most prominent thinkers of his time. He was a philosopher, naturalist, diplomat and bishop, among many things. Alongside his many endeavours, he had time to reflect on snow in his commentary on Aristotle's *Meteorologica*. He thought highly of the cold, claiming 'the north wind strengthens the virtues, the sun makes them weak', and is said to have been the first person in Europe to refer to a snowflake as hexagonal – some 1,400 years after Han Yin in China. Albertus referred to it as a star shape – *figura stellae* in Latin. It was Cambridge historian and sinologist Joseph Needham who popularized Magnus's observation. In 1961, Needham and his Chinese colleague Lu Gwei-djen published an article, the point of which was not to appreciate Albertus Magnus's achievement, but more precisely that the snowflake was a Chinese concept, and that China had been far ahead of Europe. For Needham, this was a matter of some urgency. Despite the Silk Road, the idea of the snow crystal, for all its beauty, doesn't seem to have found its way over from East to West. In addition to the compass, gunpowder and the stirrup, it turns out that China invented the snow crystal, too.

At the beginning of the fourteenth century, towards the end of the Middle Ages and at the threshold of the Renaissance, Dante Alighieri, exiled from his beloved hometown of Florence, wrote *The Divine Comedy*. In this work, snow, which Dante and his presumed readers could often see on both the Apennines and the Alps, makes several appearances. We find it first in the 'Inferno' – that is to say, in Hell. In the third of Hell's 'Circles of Torment', Dante encounters people who've committed the mortal sin of gluttony, having wallowed in food and drink during their lives on earth. Now they stand in stinking mud, whipped by eternal freezing-cold rain: 'Gross hailstones, water gray with filth, and snow / come streaking down across the shadowed air.' The gluttonous are torn apart by a dog-like beast with three barking heads, the abominable Cerberus. A little later, in the Seventh Circle, sinners who have committed violence against nature and art, among them the Sodomites, are forced to run on burning sand under a rain of fire akin to a snowfall of large silent flakes: 'as snow descends on alps when no wind blows'.

The theme of transformation recurs with peculiar force when Dante reaches 'Purgatory', where the guide Statius explains to Dante and his companion Virgil that the remarkable thing about Mount Purgatory is that, although it's on earth, it's simultaneously free from all the seasonal changes connected with earthly existence. The proof: no snow, neither hail nor hoarfrost, falls on this mountain. It continues in Paradise. 'Now, just as the sub-matter of the snow, / beneath the blows of the warm rays, is stripped / of both its former colour and its cold.' For only spirit is there, which is not subject to such vicissitudes. Again, the intent is to show that the traveller through the highest levels of Paradise is freed from all the temporalities of earthly life, just as the snow itself, already so featureless, loses what little it possesses of sensual qualities, its coldness and colour, when under the sun's rays. All that will remain of Dante's intellect is pure understanding, a blank slate, an original *prima materia*, and that's what's required for him to be able to take in the supernatural that awaits him.

Snow returns in the work's hundredth and final stanza, in the form of a parable. Dante has now seen his beloved Beatrice and been dazzled by the marvellous light of eternity emanating from God, far exceeding anything he could previously have imagined: 'such am I, for my vision almost fades / completely, yet it still distills within / my heart the sweetness that was born of it. / So is the snow, beneath the sun, unsealed.' Nothing can compare with the divine light. Anything familiar, like the snow and its consistency, has to dissipate before the incomparable. So the disintegration of the snow becomes the ultimate sign that his necessary transformation has been accomplished, that his intellect has sloughed off its burden. Whatever that really means, it's a kind of acknowledgement of this marginal and innocent medium. The snow can serve as an example of everything earthly and ephemeral that needs to be given up at this highest point of Creation, where, in a sense, all must be lost when the seeker is fully elevated into light and bliss.

This is no eulogy to snow. Mount Purgatory – which Dante climbs on his way up to his beloved Beatrice, all the way to the highest spheres, where only the angels dwell and finally God, most supreme of all – is not bedecked with any actual snow. Snow didn't have this divine status in Dante's time, and this allows its absence to make Dante's point. Snow functions here as a slight substance, whose impermanence is allowed to represent what the soul's encounter with the highest being presupposes, to be able to leave the last remaining defensive walls and the temporal in order to meet the light of love, transformed and receptive.

Dante's snow is embedded in the world of ideas. As in antiquity, more than a millennium earlier, in the art of the Middle Ages, real snow only rarely appears on the peaks of distant mountains. One such example is found in an illustration, attributed to the Dutch painter Paul Limbourg, from the illuminated *Très riches heures du Duc de Berry*, a Book of Hours from the fifteenth century, which includes more than a hundred hand-painted images of inexpressible beauty. The illustration for the month of February shows a wintry landscape with high mountain peaks in the background and a farm in the foreground, perhaps somewhere in France. Sheep crowd in a pen, a group of crows eat seeds. There's a cart, a fence, an open-gabled house, people at work, a woman driving a loaded donkey up a hill. The rooftops, the trees and the ground are all covered in snow.

Depictions of snow and snowfall came later in art, then. But in the very cold 1560s, for instance, snow is present in several paintings by the Dutch artist Pieter Bruegel the Elder. He has a silent snowfall hanging over *The Adoration of the Magi in a Winter Landscape*. The three wise men visit Jesus and the Virgin Mother, but Bruegel has set the scene in a contemporary Flemish village in snow. In *The Census at Bethlehem*, the Habsburg double eagle hangs on the exterior of an inn to show the involvement of the authorities, then in the hands of the unpopular Philip II. Meanwhile, the surrounding villagers go about their lives and the children throw snowballs. In the foreground, Joseph walks in the snow alongside the pregnant Mary riding a donkey. The most famous of the elder Bruegel's winter paintings is probably *The Hunters in the Snow*, an archetypal snow-and-ice landscape, with skaters, dogs, fires and children playing.

During the same period, Bruegel also painted a number of different versions of the Gospel of Matthew story 'The Massacre of the Innocents'. The three wise men from the East come to Jerusalem asking where they can find the newborn king of Judah, whose star has illuminated the sky. Emperor Herod, who wants to remove any possible rival for the throne, especially one from David's lineage, sends his soldiers to Bethlehem, where the newborn is prophesied to be found. Their mission is to kill every baby boy under two years of age. But Jesus was spared because an angel had warned Joseph, who flees with Mary and the baby to Egypt. In the Christian legends of the Middle Ages, the episode turns into a bloodbath, with thousands of dead children.

Once again, Bruegel alludes to a well-known biblical story and connects it to the present: the anger of the Dutch towards Spanish supremacy is widespread and preparations are underway for a revolt, which will result in a long war for independence. Bruegel interprets the mood of rebellion in a winter scene where he has the emperor's mercenaries ride into a snow-covered village in Brabant, Belgium, to conduct a massacre. These kinds of events occurred in real life, too. There's no doubt that the figures are the emperor's troops. The leader holds a staff in black and yellow, the colours of the House of Habsburg. The snow in the painting can be interpreted both as a symbol of the social climate that has reached freezing point and as a framing of the evil of power. But it can also be interpreted as the immutable position of moral conscience which, in its original childlike innocence, reminds the viewer of the crime being committed.

The biblical infanticides, which likely never took place, have given rise to what is known as the day of the 'Holy Innocents', which has been commemorated throughout the Christian world since Roman times. It is true that Bruegel created his winter paintings during some unusually cold years, but we can understand the snow in his painting better if we don't interpret it simply as a weather report. Instead, we should read it as a reminder of innocence, and how the weak in society need protection, as well as the principles of freedom and independence on which society is based, and which the Dutch recognized were under threat.

In Bruegel's time, snow was nothing unusual. He was active during the long cold spell that began in the fourteenth century and ran until the nineteenth century, dubbed the Little Ice Age by an American glaciologist long afterwards, in 1939. By then, thanks to Esmark, Agassiz and others, it had been known for a hundred years that an ice age had occurred fairly recently, ending just over ten thousand years ago. It was also increasingly clear that there had been ice ages before that.

The unusually severe winter of 1565 inspired Dutch artists in particular to depict the new kinds of scenes afforded them by frozen canals and deep snow. But even in the harshest cold, or under the harshest empires, snow was never permanent. Winter snows melted. There were also periods with little snow. This made winter scenes challenging to complete, so snow as visual inspiration was difficult to invest in.

In the nineteenth century, the Little Ice Age of Northern Europe was coming to an end, just about the same time that National Romantic snow-painting, a completely different way of thinking and talking about snow, would give snow an important place in art once again. By then, Dante and the monks of the Middle Ages had been dead for some six hundred years. In their time, snow, along with mountains and wild forests, was a part of nature that most people deeply respected and preferred to keep at a distance, and had no reason to romanticize. Bruegel had not romanticized it, either. His snow was playful, serious, human – and political.

Snow Cures Madness

*

WE GET ENGAGED IN NORTHERNMOST NORWAY, on the snow in gleaming May sunlight, near the Arctic Sea. It's something that we do: ski in the mountains. On another occasion, it's a singular kind of day – simultaneously winter, cloudy, warm and windless at a thousand feet up. We're advancing at an angle, diagonally across a steep mountainside. Visibility is poor. After a while, we're enveloped by fog. It's very slow-going. But we're going to reach the plateau on the other side any minute now, right? That's what I'm thinking. But we can't see anything. It gets steeper. I reach my hand towards the mountainside to feel the snow. It's damp. It has snowed, then thawed. Our touring skis, with their thin steel edges, are like knives in the hide of an animal that's going to be skinned. The skin will come off in a single sheet. One incision will do it.

'We have to turn around ...' I hardly dare open my mouth for fear that vibration from the sound will trigger a grain of snow that will set the entire mountainside in motion. Something the Swedish archbishop Olaus Magnus said from his exile in Rome in 1555 flies through my reptilian brain. He thought that the flap of a small bird's wings was enough to cause an avalanche – 'especially in the case of a southerly wind . . . whereby castles and entire villages have been destroyed'. Snow this wet weighs hundreds of kilos per square metre. We're going to be flattened, then pinned down, then suffocated.

It's too steep to go up – or down.

The preceding chapter was my lengthy way of saying that throughout antiquity and the Middle Ages, snow had not been a major subject in science or art. In the Bible, snow is mentioned a mere twenty-four times, mostly metaphorically. By the time people in Europe started taking a serious interest in learning more about what snow crystals look like and how they form, European scholarship had taken a somewhat casual interest in snow for almost two thousand years without really learning much about it, let alone anything new. It would certainly have been more rewarding for those who were curious to consult people who actually lived near snow. That is to say, the Nordic peoples, or people living in mountainous regions, who'd developed their understanding of snow over thousands of years and knew a great deal about it.

For them, snow was their way of life as a matter of course. This was true of the Gwich'in and various Iñupiat and Inuit peoples in Greenland, the Arctic parts of what we now call North America, and, of course, many other groups in snowy areas south of the Arctic Circle, such as the Aleuts, Tlingit, Yupik and Eyak. It's still the case in present-day Russia for the Komi, Nenets and Yakuts, the Samoyeds of the Arctic Ocean, the Chukchi of the Barents Sea and the Koryaks of Kamchatka, among many others. In the Nordic region, there are the different groups of Sámi. In Japan, the Ainu in Hokkaido. Add to this the indigenous peoples of the Andes, where people first arrived eleven or twelve thousand years ago, and at the earth's 'third pole', the Himalayas and the Pamirs, with their offshoots within the Turkic peoples in the north and the Pashtuns in the west, as well as within the Nepalese and Chinese, and all the others who have lived in snow in this part of the world. That includes the peoples of the Altai Mountains on the Russia–China border, who, according to cave paintings, may have been the very first to use skis, as far back as ten thousand years ago.

Conversations about snow were rare between these people and the people doing all the writing in Europe's towns and cities, monasteries and universities. Whole worlds of knowledge failed to interact. Deep experience of snow was invisible, and would remain so for a long time.

The modern history of snow is a long series of beginnings that have only recently come together into a more cohesive understanding. For much of history, knowledge about snow remained largely local. It was considered an element, and as a material it was used in different ways in different places on earth. It is only in our own era that snow has become a globally recognized element, in the way that air, water and minerals have long been. Ice gained this status even before snow, which for a long time was seen mostly as a peripheral phenomenon, found mostly in the Arctic and Antarctic. Ice became a universal element around the time that our contemporary understanding of climate – that it is humans who are driving the warming of the atmosphere – emerged, primarily after the Second World War. Contemporary human action has had an effect on ice everywhere: on glaciers on most continents, the Greenland Ice Sheet, Antarctica's vast continental ice sheet, and on sea ice, especially in the Arctic

Ocean. All of these ice sheets began contracting in the twentieth century, in every location – the cause was the same. Snow is now following in the footsteps of ice, its global career as an element only just beginning.

A precursor to this understanding of snow's elemental nature was the term 'cryosphere'. *Cryos* means 'cold' in classical Greek, simply because the original meaning of the word was 'ice', and ice looks like crystal – *krystallos* in Greek. The word cryosphere was coined by Polish glaciologist Antoni Dobrowolski. As a student in Warsaw, Dobrowolski was a radical nationalist who fell into disfavour with the Czarist Russian regime that controlled Poland. He was imprisoned in the Caucasus region but managed to escape and, after studying in Switzerland and Belgium, made his way to Antarctica as a member of Adrien de Gerlache's Belgian winter expedition of 1897–99. Over that winter in Antarctica, on a ship embedded in the ice and snow, surrounded by darkness and total silence, Dobrowolski thrived as he'd never done before. He loved the cold.

Dobrowolski spent some of the First World War in Sweden, both to escape the war and to study snow and ice alongside Swedish researchers including geographer Axel Hamberg at Uppsala University, who conducted research in the Sarek National Park, and whom Dobrowolski frequently cited. At the time, Dobrowolski was also drafting his monumental book *Historja naturalna lodu* ('The Natural History of Ice'), in which he considered snow and ice from all conceivable perspectives. According to his account in the book, he completed the manuscript in 1916, his last year in Sweden, though it wasn't until 1923 that it was published, in

Warsaw. In the book, he launched a new concept that would eventually encompass the entire frozen world, including ice, snow and tundra. Not only the geosphere and hydrosphere, but also atmospheric freezing, including snow formation, hail and all forms of frozen celestial matter, had a place in what Dobrowolski called the cryosphere.

The term 'cryosphere' drifted about for a long time without a real foothold in scientific literature, probably partly because most scientists couldn't read Dobrowolski's work in its original Polish. Over time, however, the work became increasingly admired, as its contents slowly but surely spread throughout the scientific community. For a long time, it was unrivalled as an overview of international research. Nonetheless, most researchers were cautious of the *cryosphere* concept, and even more so of Dobrowolski's attempt to launch 'cryology' as a common term for all research within the realm of coldness, in all its dimensions. 'Cry' was a word invoking confusion and weakness, explained the interwar snow authority Gerald Seligman, whose great and literally magisterial book on snow, *Snow Structure and Ski Fields* we encountered earlier (and will encounter again).

Dobrowolski was devastated. Soviet scientists, however, picked up on the word and a textbook on *geokriologiya* was published in 1940. The Soviets were also more receptive to the idea of a unified cryosphere, which gained traction in the Soviet Union in the 1950s. The big international breakthrough came in the 1970s, after Dobrowolski himself had been dead for a couple of decades. By then, satellites had begun orbiting the earth, offering an overview of all the snow and ice that actually existed, and Dobrowolski's concept suddenly took hold, silencing all doubters.

The cryosphere is a planetary concept. And all three of its continental elements of ice, snow and permafrost are shrinking due to climate change. Snow is the least permanent of the three. There are large areas in which snow is found only in winter, and in some winters hardly at all. Snow, too, is becoming universal – not because it's becoming more widespread, but because it's part of the emerging narrative of the planet as an Anthropocene world, in which human behaviour increasingly controls even processes that were previously perceived as being purely natural. The universal snow is also *humanity's* snow, as more and more people miss the snow that used to exist.

Snow has started to concern all of us. In the past, the many attempts to understand it revolved around local and national needs – to better avoid avalanches when railways and roads were built through the Rocky Mountains, the Alps, Tibet or the Urals; to be able to measure and predict snow depth so that spring flood intensity could be gauged and the energy supply and irrigation of downstream farms was at stake. Many of these needs to measure snow arose in the nineteenth and twentieth centuries, driving the acceleration of knowledge. Snow and ice research stations were established in several countries, those in Switzerland and Japan among the most prominent. Laboratories were set up in the USA and the Soviet Union. While the Second World War and the subsequent Cold War were a heyday for snow research, many facilities had been established earlier.

We can see each new investigative beginning as a trail in the snow. The tracks lead us forward, with small movements in different directions, which soon begin to intersect, converge, and form ever-wider paths that start to point more and more in the same direction. A pattern.

The first beginnings were fairly small in scale – they were to do with crystals. The people who started to really explore snow more systematically in the seventeenth and eighteenth centuries were contending with very varied snow conditions. But no matter what the snow looked like in different places, most researchers soon realized that the crystals that made up the snow always looked pretty much the same. They had six arms (or points, sides or corners – they're referred to by many terms) and almost 100 per cent symmetry. They also exhibited microscopic variations. Maybe that was a natural law of some kind – that no two snow crystals were ever exactly alike? It seemed planned, as if some kind of creator had left their signature within each snowflake, as a small sign to humanity: 'Look, I could have made all these little crystals the same. But I can also make them different in very small ways.' This became a popular idea and gradually merged with another – one that had fascinated scientists from the start: that the symmetry of the crystal was – that's right – beautiful.

One person who started propagating this idea was seventeenth-century Prague-based astronomer Johannes Kepler. He devoted himself primarily to the movement of the planets, but was interested in most things having to do with the heavens. On New Year's Day 1611, having observed snowflakes falling during his walks through town over a period of time, he wrote a short treatise entitled *De nive sexangula* – 'On the Six-Cornered Snowflake', just thirty pages long. He then gave it to his patron, the German diplomat Johannes

Mattheus Wacker von Wackenfels, as a New Year's present. The gift was intended to be small and to represent the void, or the 'nothingness' that Kepler knew his patron admired, including in the form of the poem *'De nihilo'*, which had been written by the learned French satirist, poet and Latin scholar Jean Passerat for his own patron, also as a New Year's gift. The snowflake fitted well into that context, especially as the Latin for snow, *nix,* sounds similar to the German word *nichts* – which means 'nothing'.

In his essay, Kepler can see with his own eyes, like many others before him, that a snowflake has six sides, and he wants to understand where this six-pointed structure comes from. It can't possibly come from within the snowflake itself, he reasons. Each animal and plant possesses an inner life-force, a soul, which helps to determine its shape, but Kepler can't conceive of snowflakes having souls. That would be 'utterly absurd', he writes. There must be some external force that creates these two-dimensional stars. But what kind of force?

He entertains various suggestions. He thinks of the hexagonal cells in a beehive, which conserve both labour and space. It makes sense that God would put such architectural features into the bees he created. But why would it be needed in snowflakes? Snowflakes don't crowd each other, nor do they have work to do . . . But maybe there's something to be gleaned from the rhomboid shape of pomegranate seeds and their hexagonal arrangement within the fruit? The pomegranate might even have a soul, Kepler says, but that can't be the whole explanation. Comparison with the pomegranate gets him nowhere. In the plant kingdom, he reflects, pentagons, five-sided shapes, are everywhere, but why do snowflakes insist on having *six* corners? As a mathematician, he thinks of the golden ratio, which features the number five, as well as the Fibonacci sequence. But even these astute investigations don't help him understand why the snowflakes don't have five corners, but insist on having *six*? And why doesn't snow simply form itself into regular spheres? Rain spontaneously forms into drops. Why should the humidity of the air suddenly become a two-dimensional surface when the three-dimensional droplets freeze and turn into snow? Is there something in the cold itself that comprises the active force? If so, what could that be? Might there be an overarching formative principle built into Creation that necessitates each object having the appropriate shape for its task? That seems to be the case for animals and plants. But what could be the principle behind a hexagon, whose flat, wide shape doesn't even enhance its own stability?

Could it be that the liquid's encounter with the cold is determined by properties of the liquid? If, for example, salt is dissolved in the water prior to freezing, would the flake take on a different shape? Here, with this very thought, Kepler actually gets closer to a conceivable answer, but for now he'll have to refer to chemists, who are better able to say what the salt implies for the properties of water, and what it is that might explain the formation of hexagons. After all, Kepler wonders, maybe the Creator just decided that triangles and pentagons should be for fertile entities, while hexagons should be for those that didn't reproduce (had he already forgotten about the beehive?).

And so it starts snowing again. Kepler once more studies the snowflakes that land on his sleeve. Some form large, thick, woolly balls. But when he looks at them he discovers that these, too, are composed of flat hexagons. He gives up, praising Wacker von Wackenfels for his admiration of Epicurus, who worshipped nothingness. Wackenfels must have been happier thinking about nothingness than poor Kepler was, speculating in vain about that which cannot be known.

Kepler wasn't the first to take an interest in snowflakes, even within Europe. Albertus Magnus knew about them, which, when this was heard in China, occasioned delighted gloating that Europeans, 1,400 years after China, had only just figured out what a snowflake looks like! But at the time Kepler was writing his text, at New Year 1611, the first microscopes had just become available. Kepler himself used lenses in his telescopes and wrote a whole book on optics, so he would certainly have known how to study the snow with an enhanced eye.

The same was true of the French philosopher René Descartes, who discussed snow, 'De la neige', in one of his most famous works, Discours de la méthode ('Discourse on Method'), published in 1637. In three places, Descartes illustrated his treatise on snow with drawings of snowflakes, each, interestingly enough, having a distinctly different appearance. Other parts of the Discourse deal with geometry and optics, fundamental aspects in describing the world, among other things. It's in this same spirit that we can understand Descartes' interest in snow. Snow belongs to the order of the world. What happens in the atmosphere is no different from optics. The sky can be interpreted with physics, and what is above the earth follows the same laws of nature that apply down here. This applies, as Kepler would have agreed, to the distant heavenly bodies, as well as to the snowflakes in the air. Descartes was a keen observer, able to distinguish between observation on the one hand, and, on the other, the imaginative kind of thinking that he used to make sense of and understand the world in contexts other than what he observed with his senses. His writings are richly detailed and difficult to interpret, but bear witness to a person who really sought to understand what he observed in the world around him.

Descartes sees important differences between different manifestations of snow. Like this one, from a series of observations he made in Amsterdam in February of 1635:

> . . . the particles of ice of no particular shape assured me that the storm clouds were not always composed of little clumps or *pelotons*, but that it was only some of them that were made of randomly mixed threads. As to the causes of the falling of these snowflakes: the violence of the wind, which continued all day, made it perfectly clear to me; for I judged that it could easily rearrange and break up the layers which they [the snowflakes] formed, after it [the wind] had formed them; and that, as soon as they were thus rearranged, they could, with one of their edges inclined to the earth, easily penetrate the air, being rather flat and heavy enough to travel downwards.

Peloton can mean either bullet, platoon or bunch. Here it's a matter of six snowflakes arranged around another one, making a total of seven. In the same passage, Descartes tells us about another specialized form of snowflake that later research would give its own name, *tsuzumi* – crystals bound together with a link of ice, forming a double crystal.

The fact that Descartes studied crystals only makes complete sense if you think of them as being part of the larger cosmology, or a Creation story in which everything in the world came into being. Somewhat like in the Bible, but with noteworthy departures. Most prominent among them is that, according to Descartes, Creation took much longer, not just six days of work, as in the Book of Genesis. And he makes a lot of additions about how stars and planets were formed, and how the oceans came into being. His world is much more dynamic than the one in the Bible. Everything in this Creation had meaning, even a snowflake. Snowflakes belonged to what Aristotle had already called meteors – that is, things whose place is in the air. Descartes claimed that if only the scientist had a 'method', the reality behind these phenomena could be unravelled. Descartes didn't need to be the one to solve the riddle; he just showed that it would be possible one day. His was one of the first forays into the contemporary history of snow.

Another initial trail in the snow comes out of Denmark, but is at the same time pan-European and shows how the threads of crystal are linked between environments of the most diverse kinds across the continent. Thomas Bartholin, a true polymath, started out as a theologian, then went into law, then studied mathematics, languages and history, before settling on medicine and getting his medical degree in the Netherlands. This was followed by a doctoral degree in Basel. He travelled extensively, studying at the universities of Leiden, Paris, Montpellier, Padua and Naples, and immersing himself in botany in Messina, Sicily. Once back in Copenhagen in 1647, Bartholin became a professor of philology, then the following year a professor of anatomy, and in 1661 published a work entitled *De nivis usu medico obser-vationes varia* – 'Various Observations on the Medical Use of Snow'. The book delivers what it promises – an overview of different ways of using snow in a medical context – and is filled with learned references and lofty aspirations. Snow soothes fevers, calms palpitations, cures an upset stomach; it also quietens the agitated and confused – *Nix delirium sedat* – snow calms madness. Bartholin invokes the ancient physicians Galen and Hippocrates, philosophers such as Anaxagoras and Seneca, and a range of contemporary medical authorities.

The writing would be most notable for its contention that a mixture of snow and ice could provide an effective anaesthetic for surgical procedures. Not a completely novel idea, as medieval Arab physicians had used it, and the method was described by the philosopher and physician Avicenna in the early eleventh century. But since that time, the topic hadn't been taken up much in medical literature. In his book, Bartholin instructs the reader on how to apply snow and ice to the surface of the body area to be operated on, in parallel lines laid out at a distance from each other so as to avoid frostbite. After fifteen minutes, the incision is made and causes the patient no pain.

Never before had such a catalogue been presented concerning the medical uses of snow and, generally speaking, there's no contemporary equivalent, either. Bartholin had no reason to be dissatisfied with his work, but makes no claim to originality for the anaesthetic technique, admitting to having learned it from a doctor in Naples while he was living there for a few years in the 1640s.

On the other hand, the crystallographic studies by

Thomas Bartholin's younger brother Erasmus were definitely original. They extrapolated from the work of Kepler and Descartes, and were soon widely known, but Erasmus certainly didn't stop with the substance of snow. He was a professor of both geometry and medicine, and he, like his brother Thomas, had many and various interests. He was deeply fascinated by minerals, and delved even deeper into the subject of snow in a special appendix to his brother's work, *De figura nivis*, which concerns the shape of snow, which Erasmus describes at the most detailed and microscopic level. His interest in snow lies in its crystals. Teeny, tiny structures that, as they pile up in large quantities, build the volatile and fluffy substance that has so many strange and useful properties.

Possibly the most significant way to understand the importance of the Danish brothers' achievement is to consider the famous work *Micrographia*, published in London in 1665, written by Robert Hooke, at that time Curator by Office and in charge of experiments at the Royal Society, which had been founded only a few years earlier. It was the first major attempt to show what the world looked like at the smallest scale, using a microscope that Hooke himself had helped to design. The book became a sensation and a bestseller, with new editions constantly being published. It includes the now-famous image of a louse, a pest familiar to most people at the time. In *Micrographia*, this vermin is rendered at a gigantic scale – on a fold-out sheet that quadruples the already impressively sized volume. The louse is seen from below, in all its fascinating and arguably repulsive details; somewhat hairy, with sharp claws at the ends of its armoured legs. The same book featured pictures of snowflakes. They, too, caused a stir, but of a more enjoyable kind. Snow was at least a little more pleasant than lice, even if it was cold and wet most of the time. Here it could be seen to be beautiful, too, perfectly formed in the shape of a star. Regular, but with each flake always having some small distinguishing detail differentiating it from all the other 'starry flakes'. No one had seen snow in such sharp detail before.

It was the *scale* that made a difference – to be able to get closer, to see more. For most people, this was snow that was totally new to them, just as the louse had been new. The latter was horrible to see magnified to such a degree, while snow seen in exaggerated size was, on the contrary, dazzlingly beautiful. This is a clear illustration of how new forms of knowledge often have completely different sides to them and are met with completely different reactions. In this case, snow and snowflakes came out on the winning side. Hooke and the Royal Society could bask in the glory of the new instrument's crowd-pleasing success, never mind that the microscope itself wasn't of British origin, but had instead been invented in the Netherlands and Italy decades earlier. New beginnings, but also new connections.

Incidentally, aren't those images in *Micrographia* rather similar to those in the Bartholins' book? Maybe they are. It's been suggested that Hooke stole them from the work of the Danish brothers. But other scholars have made credible arguments to the contrary, so we may as well trust him. The seventeenth century also held a different view of scientific originality and ownership of intellectual property, so there's no need to moralize here. Perhaps it illustrates how science moves forward, through precisely such borrowing, creative reworking, and new application.

And there was something else new that was now undeniably in motion. From anaesthetics in Naples, through medicine and crystallography in Copenhagen, to microscopy in the Netherlands and London, all in the space of a few decades, a whole new picture of snow would emerge, far removed from the Sámi, Yakut, Inuit, Ainu and other peoples who lived with and in the snow. In Oxford in 1665, the naturalist Robert Boyle had published a powerful 880-page work called *New Experiments and Observations Touching Cold*. Similar experiments had been carried out by the duchess and writer Margaret Cavendish, one of the few female experimentalists at the time, though these two came to very different conclusions. Their works focused on research on snow, not individual crystals (although Boyle returned to this in his work on gemstones in 1672). This demonstrates an important point: that snow was the subject of growing interest.

Boyle was the most influential and innovative experimentalist of his era. According to the Royal Society, the subjects of cold, ice and snow had previously been 'almost totally neglected'. Now, with Boyle's book, Cavendish's research and Hooke's microscope, they'd proved to be the opposite – both highly noteworthy and fruitful. In Sweden, too, the topic attracted attention. In 1761, the physicist Johan Carl Wilcke reported to the Academy of Sciences on his experiments with 'snow forms' – the shapes of snow crystals. He refers to Kepler's writing of 1611, in which he says that every winter since that time further images had been made, but still just as little was known about the forces at work in them. 'We can collect fully formed snowflakes, but can only guess at the reason for them.' Wilcke's experiments involved mixing water made from melted snow with various types of soap and cleaning agents, which he then blew with a smoking pipe into cold winter air that he simply let in through the window. (As yet there wasn't a way to create cold artificially.) In the bubbles, Wilcke saw wonderfully symmetrical and – often, but not always – hexagonal shapes forming very quickly and with breathtaking precision. He was able to make out an immense variety, but also their uniformity. So, much the same as his predecessors had. This is as far as Wilcke gets. He laments that he's unable to go up among the clouds to try to find out what's really happening.

Wilcke's terminology is interesting. Nowhere in the text does he use the words 'snow star' or 'crystal'. It's as if all this terminology from the European debate failed to interest him. Otherwise it would have been natural for him to use the same terminology – he'd seen the star pattern, after all. However, Wilcke returned in 1769 with *Nya rön om vattnets frysning til snölike is-figurer* ('New Findings About the Freezing of Water into Snow-like Ice Forms'), looking at the various crystals within the snowflakes. The new experiment that he described was conducted during an especially cold period in Stockholm that same winter. In the experiment, he cooled snow water and put it in round glass containers, then added small objects around which crystallization took place. Here he once again experiences that lightning-fast transformation when the hexagonal patterns appear, just as they had in the soap bubbles in 1761. The ice forms thus become 'snow-like'. He even suggests that he has succeeded in simulating the crystallization that happens when snow forms occur in real life, up there among the clouds in the 'air sphere'. Wilcke has discovered something important, and he's witnessed it with his own eyes.

His text brims with the experimentalist's self-confidence. At this point he is clearly much more enamoured of the beauty of the resulting snow and ice forms – and he uses the term 'snow stars'.

White Magic

*

WHY AM I SO RECEPTIVE TO SNOWFLAKES? Why do I melt when they meet my soul? Why am I so curious about how these frozen drops from the atmosphere intersect with our lives?

Because I believe that we humans are fragile. We need a context that helps us be a little better. We have to manage our reassessment and transformation.

The simplest words take on a new meaning.

Snow is one such word. We are violent towards snow yet at an individual level we 'melt' at the beauty of snow. Snow equally can be violent to us, but it isn't against us. The water-heaven gives birth to, bears down to us, both an act of grace and an affliction.

The disappearing of snow is something that will have been our own doing. Not a feature of Creation or a consequence of the arrival of spring. An *act*, not an event. And therefore not an act of grace, either.

The people who've created this loss are mostly the more or less affluent people of the world. We are in the process of eliminating acts of grace from the world. This poses a great danger. Without these acts of grace, the world is worth nothing more than what we humans have done to it, all of which is perishable. We need to discover what is eternal in order to discover ourselves and the responsibility we have to preserve the very meaning of our existence on Earth.

It's not the forces of nature that are against us. We understand them well enough already. And it's not a paucity of knowledge about nature that's preventing us from taking action, it's societal forces. Violence is exercised not only over distances in space, but also through distances in time. Our descendants will suffer. Some people living now give no thought to the people who'll come after us. They seem to put their faith blindly in a miracle, but never mention what that miracle is going to be. They might mention 'progress' from time to time, but they don't seem to realize that the very word presumes that there's time, and it's our descendants who'll evaluate the results of this 'progress'.

It was a beginning. It consisted of several new beginnings that would prove to be interconnected. These collective beginnings, extending over a couple of centuries, would prove to be absolutely decisive, both for the snow we'd get and for the further new beginnings that would follow. Snow was always held to be among the *things that could be understood*. We just didn't understand it *yet*. Wilcke's efforts had by no means solved the riddle. He had made snow more visually accessible, but he had no better idea than anyone else what actually happened to create snow.

At that time, atoms and molecules were still unknown, at least in the modern sense that we understand them. What was new, however, was that images of snow could begin to circulate in society, through Hooke's microscopic views and the attendant snow taxonomies – ways of describing them. The components of snow became public property *as images* long before they became accessible as physical objects for in-depth analysis. The fact that most people had to make do with Hooke's rather free artistic rendering for a while probably didn't matter that much. People in Europe and the rest of the world were starting to know what a snow crystal was. The snowball might not yet be rolling, but at least the snowflake was slowly wafting its way down and landing as a silent apparition, even if more visual than tangible.

Snow in 1800 was a very different creature to snow in 1600. As evidence of how established snow had become, consider William Scoresby, a versatile British whaler and explorer of the North Atlantic and Arctic in the early decades of the nineteenth century. Scoresby had studied snow stars and snowflakes already, while in his teens. Over time, he compiled a comprehensive taxonomy of them, organizing them into major groups and naming the different varieties. Scoresby's Arctic research showed that the snowflakes in the far north were even more symmetrical than those falling further south. He also went further than his predecessors in his appreciation of the aesthetics of snowflakes. At just over thirty years of age, he published a two-volume work on his Arctic research called *An Account of the Arctic Regions* (1820), from which the following is often quoted:

> The extreme beauty and endless variety of the microscopic objects procured in the animal and vegetable kingdoms, are perhaps fully equalled, if not surpassed, in both particulars of beauty and variety, by the crystals of snow.

Those are the words of a scientist; his predecessors would have sounded much the same. But it's important to read further in the same passage to find the words that provide a clue to the theme of the magic of snow, which from about this time, the early nineteenth century, begins to emerge more and more clearly:

> The principal configurations are the stelliform and hexagonal ... Some of the general varieties in the figures of the crystals may be referred to the temperature of the air; but the particular and endless modifications of similar classes of crystals can only be referred to the pleasure of the great First Cause, whose works, even the most minute and evanescent, bear the impress of His own hand, and display to his intelligent creatures his vast and beneficent wisdom.

Here is a reference to God, as the 'great First Cause'. A few years after writing these words, Scoresby took up theological studies at Cambridge, and for the rest of his life alternated between research voyages in the North Atlantic and Arctic Ocean and his ministry in various parts of England. Taxonomy can be seen as the more conventionally scientific side of his interest in snow, where he came up with as many as ninety-six different types of snowflake. He was by no means the first. The German physician Friderich Martens, having travelled to the island of Spitsbergen in 1671, had in 1675 already proposed a classification into twenty-four categories.

Another attempt at the morphology of snow was made by the philosopher and mathematician Donato Rossetti in Turin with his book *La figura della neve* ('The Figure of Snow') in 1681. The Great First Cause was of course omnipresent in the thoughts of these early snow researchers. But the sentiment around the subject, often religiously inflected, increased during the eighteenth century and is also noticeable in Scoresby's work. In time, it developed into what came to be called 'natural theology'. Nature became a ubiquitous example of God's omnipotence and goodness.

In natural theology, snowflakes began to appear as an increasingly popular prop. In the late 1770s, the Dutchman Johannes Florentius Martinet published a four-volume work entitled *Catechism of Nature*. Martinet was a preacher and amateur biologist who boasted of having tamed a stork and even tried to make it swallow a fish tail-first (without success). With this opus, he made a name for himself as a spiritual nature guide and educational counsellor. In the book, a little boy is led around different landscapes by his teacher and helped by this earthly shepherd to recognize the traces of the wonderful divinity in every detail of his Creation. Not the least of these were the enigmatically symmetrical snow-flakes, which were presented in the form of a special poster with oversimplified but surely convincing illustrations.

The genre was effective and proved long-lasting – Jean-Jacques Rousseau's 1762 Bildungsroman, *Émile*, is the masterpiece – and was part of what author Marleen de Vries has called the 'pedagogical enlightenment' in the Netherlands at that time. Education determined the future of societies. Rousseau believed that if children were properly educated, there would be no need for prisons. Such was the power of Enlightenment, more and more people contended. They shared a fascination with nature. But unlike

Rousseau, Martinet did not emphasize the freedom and barrier-breaking that had characterized Émile's upbringing, but instead focused on piety. Martinet followed up with an abridged edition for children in a compendium volume (*Kleine Katechismus*), which was a great success and went on to be translated into several languages. His work resonated particularly strongly in Japan, where the snowflakes had also generated excitement and were taken up by artists. One of these was Doi Toshitsura, who'd already been drawing snowflakes for some time. In 1832 Toshitsura published a work entitled *Sekka zusetsu* – 'The Book of Snowflakes' – in which he apparently collected eighty-six of his own snowflake images and twelve of Martinet's. There's some confusion as to their provenance. Maybe all the pictures were done while Toshitsura was in charge of the book's production, in the Netherlands. The images were in turn reproduced by Bokushi Suzuki in his 1837 book *Hokuetsu seppu* – 'Snow Stories from the Northern Etsu Province'. This was widely circulated and marked the start of a period of research on snow and romance around snow in Japan, with consequences few could have imagined. It's the initial seed of an explanation as to why Japanese research on snow is so prominent in our own time. More on this shortly.

Here it may be appropriate to leave the seekers of crystals and their new beginnings for a moment. The multiplicity, as you see, is already starting to be substantial. The tracks in the snow spread out in different directions, but they also intersect all the while, sometimes merging to create wider paths, some of which lead forward. We're getting more glimpses of one of these wider tracks. We could call it a path of emotions, but it's also a romantic trail, with elements of reverence for what may not be comprehensible through science alone.

Scientists themselves had reverential moments. In December 1859, when Irish physicist John Tyndall was on the Mer de Glace overlooking Chamonix, he was struck by an explosion of beauty from the sky, in the form of snowfall:

> Some time afterwards the air became quite still, and the snow underwent a wonderful change. Frozen flowers similar to those I had observed on Monte Rosa fell in myriads. For a long time the flakes were wholly composed of the exquisite blossoms entangled together. On the surface of my woollen dress they were soft as down; the snow itself on which they fell seemed covered by a layer of down, while my coat was completely spangled with six-rayed stars.

This is what Tyndall wrote in *The Glaciers of the Alps*, published the following year. He can't reconcile this experience with the widespread belief that Creation exists for human beings, that the beauty of meadow lilies exists to satisfy our very understanding of what constitutes beauty. If that's the case, how can the snow be allowed to fall in such an inaccessible place, with all of the beauty never seen by anyone? 'Nature rained down beauty', he recalled of snowfall during an expedition to the Swiss Mer de Glace, and it had done since the dawn of time without a single human being present. 'Whence those frozen blossoms? Why for aeons wasted?' The questions are ultimately addressed to the Creator of everything. But they are also rhetorical. In fact, they are a critique of the doubters. Nature understands what everything is for. Nothing goes to waste. There is meaning in abundance. There is a bigger plan – it's just that *we* don't understand it yet.

Another sign that such an interest was in the *zeitgeist* can be seen in the form of a minister's wife in Portland, Maine. Here, Frances Chickering documented more than two hundred different snowflakes in the mid-1800s, in a somewhat unique way. She first captured a snowflake on a piece of dark cloth, held a powerful magnifying glass over it, and then quickly cut out the actual shape from paper. From this industrious creation the Chickerings' clever neighbour made lithographs that were bound together in 1864 into a book, *Cloud Crystals: A Snow-Flake Album*. Seven snowflakes per lithograph, all on a burgundy background against which the white of the crystal stood out with fluorescent magic. The publication was a success in the growing book market revolving around the family and its values. Around the kerosene lamp the images of snow were to be savoured and contemplated, at once captivating to the eye and edifying for the soul. It all began, Chickering herself wrote, in a chance observation she made of 'the beauty of a snowflake on a dark windowsill'. The book had no named author; as a clergyman's wife, she was perhaps content to be credited as 'A Lady'. But Chickering writes the text herself, in educated and enthusiastic fashion, in addition to the scholarly texts she collected on the subject of snow, along with extracts from the works of Shakespeare, John Ruskin, Ralph Waldo Emerson, and other names in the Anglo-American canon of the time.

Long the pastime of astronomers and philosophers, by the mid-1800s snowflakes are taking up a wider place in the public imagination. New groups are taking notice:

naturalists, doctors, priests, self-taught amateurs. They attract new audiences in a wide variety of settings. Some of these circles were undeniably religious ones.

In New England, the Reverend Israel Perkins Warren (who, by the way, ends his days in Portland) is also active. He sees the potential of snow as something that could promote piety. His book *Snow-flakes: A Chapter from the Book of Nature*, from 1863, is reminiscent of Johannes Florentius Martinet's in that he posits that snowflakes are signs from God. But Warren's approach is more direct, rousing, in a Sunday-school kind of way. Warren holds from the outset that snowflakes are so beautiful and complicated that they can only have been created by God. They almost constitute *proof* of God. It is, he says, 'legitimate to ask, What but an infinite power could muster the capacity and instruments needed to fashion even a single snowflake? The tiny entity says, as you examine it – "The hand that made me is divine"'.

With Warren, the language becomes more expansive than ever before, invoking precious stones and metals. Gold dust flutters above his words, and the clouds sparkle with diamonds, almost like the Book of Revelation. Each aspect of the snowflake has a concrete message. Its variety of shapes points to God's Creation as representing 'unity in diversity'. Its perfect form conveys the meaning: 'Obey God. His laws for the snowflake are meant to make it beautiful and useful. So are his laws for you.' The snowflake's message is the source for the titles of the book's chapters: 'Perfection', 'Purity', 'Grace', 'Beauty', 'Weakness', 'Joy', 'Charity', and so on.

From this point, many threads lead forward to the winters and white, normative snow of the modern West; even to

Disney's snow-covered Christmas parades and 'If you keep on believing / The dream that you wish will come true'.

Tracks have now been drawn in the white snow. Several ran inwards, towards the magic of the soul and the imagination of the people. Others penetrated even further into the crystals of physics. The tracks will intersect. And it won't be long before someone realizes that you can take a box camera, go outside and, if it's cold enough, photograph individual snowflakes. See what they actually look like.

There are new beginnings afoot.

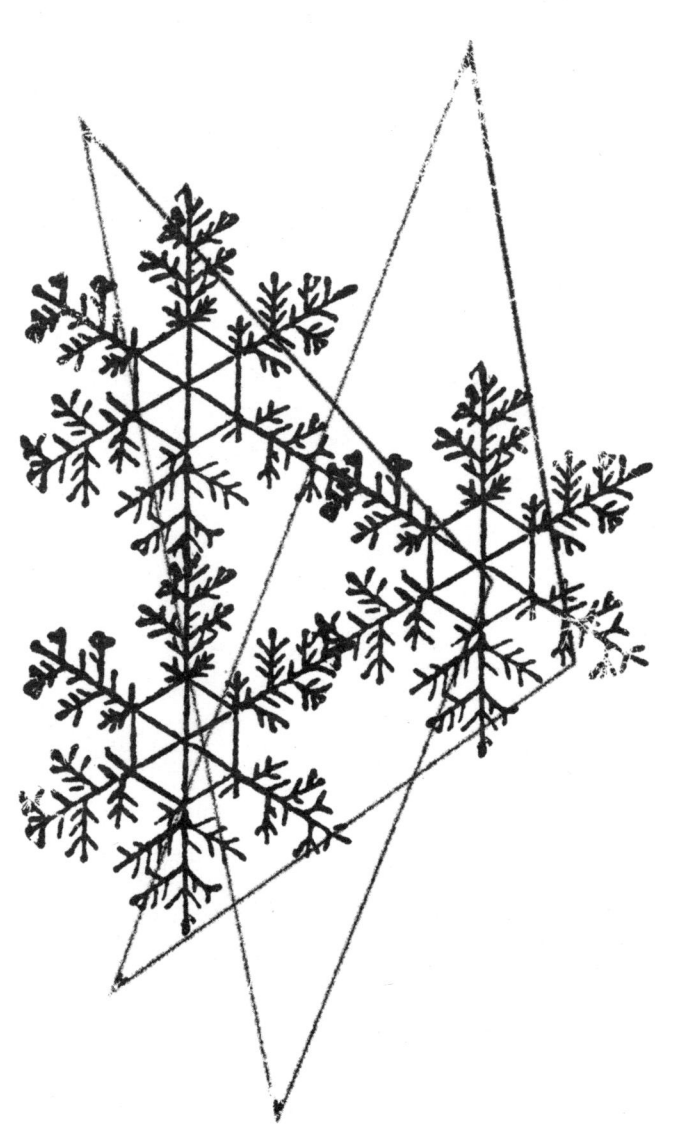

Letters from Heaven

*

IT'S SEPTEMBER 2017. Some photos are being exhibited at the Kunstnernes Hus art gallery in the middle of Oslo. They were taken in the 1930s in a laboratory, on the island of Hokkaido in northern Japan, by physicist Ukichiro Nakaya, who had managed to create snow. The images have been given new life now, in this era of snow's disappearance.

Our snow is disappearing. Nakaya made his from scratch.

It's those artificial snow crystals of his that I'm getting to see today. There are speeches and discussions throughout the entire day. It's strange how many people there are. Can snow really be that appealing? I speak on the subject 'The Political Aesthetics of Snow and Ice', insisting, as I've started doing, that snow is a political concept.

At dusk, a few streets away, we're guided up through the concrete skeleton of the building that will become Norway's new National Museum. The atmosphere is one of anticipation. On the way up, guards in safety vests guide the guests so that we don't fall down any concrete shafts. It's been raining, and the air is humid. Perfect for snow formation, I think. If not for the fact that it's still the mild weather of early autumn.

Up on the roof, the physicist's daughter, eighty-four-year-old artist Fujiko Nakaya, gives a performance-art piece featuring the 'fog sculptures' for which she is famous. An ephemeral medium, just as her father's snowflakes were. There is also an evocative synthesizer performance by composer Ryuichi Sakamoto, while two traditionally clad Japanese dancers stand out in relief against the backdrop of the Oslo fjord.

Everything is dizzying. Beyond the dancers and beyond the fog lies the grand Akershus Fortress. Beyond that, behind the forest and mountains, lies Sweden. In the other direction, more forest and mountains, beyond which lies the Atlantic. Surrounding me is the current moment of art; below me, the promise of a greater future for art.

In this moment, everything has come together. With the help of the elements. The spheres. The generations. The cultures. Physics. Art. History. Loss. Our plight. And the snow, where it all started and ended.

I realize then that it is possible, necessary even, to see all of this in context. It occurs to me again that snow is the great subject. The one that I won't be able to escape. The thing that has already captured me, without my really knowing what to do with it.

Since that evening, the National Museum has been completed. I've been able to see the works of Harald Sohlberg and Edvard Munch again with my own eyes, and also contemporary art, featuring snow, by the Sámi artist Britta Marakatt-Labba. Art by both Sámi and non-Sámi, Norwegians and non-Norwegians, in a wise expansion of the museum's remit from the Norwegian to the Nordic to the universal.

What I became aware of most of all at that moment on

the museum's roof was that there was something connecting wildly divergent points in our world. At first glance, they didn't seem to belong together. They had been brought together by forces beyond my control. That was what moved me. I've been trying to understand why I'm so susceptible to strong and sudden emotional reactions. These happen when the insight comes to me that there has been an autonomous bringing-together of things that were not meant to belong together. Not through my own active thinking. Or maybe this is what it means to *think* . . .? The striking thing is the feeling. As if I want to honour the insight because I don't feel it's of my own doing. I call it grace. I think it belongs to the world, this possibility of creating new meaning by bringing things together.

Epiphany. I don't imagine that epiphany needs to be something supernatural. Everything it involves can be understood within the framework of what can be accounted for, probably right down to the last quark, for those who want to do so. In my experience, this is about being humble. I have to let myself be small and let the experiences I've had, in whatever ways I've had them, come together. To some extent, surrender to them.

I remember afternoons in January 2012 on the windy beaches of the peninsula south of Cape Town. Lying there in the strong wind and the coarse, rust-coloured sand, reading Thomas Mann's 1924 novel *The Magic Mountain*. I noticed that the protagonist, engineer Hans Castorp, had himself seen snow under a microscope. He knew what precious things it comprised: 'the exquisite precision of form displayed by these little jewels, insignia, orders, agraffes – no jeweller, however skilled, could do finer, more minute work'. Why am I shivering at this thought? I wondered. It can't just be the cool air and my damp towel. Diamonds again. Falling gold dust.

During a skiing trip on the mountain, Castorp experiences an epiphany: he witnesses the transformation of the human interior by snow. Thomas Mann is a contemporary of Nakaya. His time period is after the Great War. The next great war is still unknown. What does this snow have to do with me, I ask? With my mother's glittering eyes, the snowdrifts of my school breaks, my devotion to the falling snow? It's not me who's catching up with the snow. It's the snow that's catching up with me.

This is what I realize on the roof of Norway's great new museum. And I believe that every human life is like this – a unique version of the total world context that we simultaneously share. Not unlike the snowflake itself. The snow can tell us about our own destiny. In other words: a story. It was there all along. It just needed to crystallize.

It took several years before Nakaya managed to make his first artificial snowflake. Before that, he had studied natural snowflakes. He called them 'letters from heaven', an idea that is taken one step further by Orhan Pamuk, who in his novel *Snow* speaks of a 'hidden symmetry' of the world. It consists of each individual having his or her own snowflake, which is their personal destiny, an idea that comes from Nakaya. The snowflake is a command from the Creator of everything. All of us, the whole of humanity, are the snow that falls and, as Pamuk writes, throws 'a veil over hatreds, greed and wrath'.

We can snow over evil.

'Everyone else around here hated winter, but I was overjoyed . . . from the very beginning, it was the snowflakes that fascinated me the most.' This statement could have been made almost anywhere that there's snow. It could have been made by me, in my room in the house by the River Ångermanälven, or by thousands of other people. This particular iteration comes from Wilson A. Bentley, who grew up in the small town of Jericho, Vermont, in the far-northeastern United States near the Canadian border. A snowy place with an obligingly biblical name.

Bentley had good reason for saying such a thing. He was truly fascinated by snow, and when, as a teenager, he was given a microscope by his mother as a present, he was able to start looking at his beloved snowflakes in sharp detail. Like so many before him, he was captivated by their symmetry and beauty. But not even Bentley could make the snowflakes linger on the piece of soft canvas where he'd trap them. If they didn't simply blow away or get melted by Bentley's body heat, the snowflakes would sublimate, vaporize and disappear. As snow does in cold air, imperceptibly, without even first becoming liquid.

Bentley struggled for a long time trying to draw the snow crystals, but no matter how much he hurried, it took too long. If he wanted to capture the way they looked, and be able to study them properly, he had to reproduce them more quickly. Luckily, photography was now widely available. At the age of fifteen, he bought a box camera and started experimenting with ways to get the snowflake into the viewfinder of his Bausch & Lomb microscope (at that time photographic technology was only just starting to be linked with microscopy). Bentley's camera was usually kept in the barn, which wasn't heated in winter, as the cold was a prerequisite. When it snowed, preferably without too strong a wind, Bentley would hold out a tray, about a square foot in size and covered with black canvas, and let a substantial quantity of snow-flakes land on it. He'd bring his crop of flakes to the shelter of the barn and quickly examine them with a magnifying glass to select an interesting specimen. Once he had found the perfect snowflake, Bentley used a small wooden pick to lift it on to a glass disc and nudged it into place with a goose feather. Then there was a final check to see if the snowflake had enough interesting features to be photographed. If not, he had to start all over again and gather new candidates. The farm had no electricity, so Bentley instead let in natural daylight, preferably in overcast weather, through a window, which gave the images a shiny, silvery lustre. Exposure times were between eight seconds and two minutes, the preparation for taking the images extremely thorough. But for the most part, his attempts failed.

Bentley wasn't alone. Other photographers developed similar methods in other places throughout the world. The Russian A. A. Sigson started photographing snow stars as early as 1872 and first published pictures of them in 1875, though he wasn't necessarily the first. The young Swede Gustaf Nordenskiöld, son of the more famous Adolf Erik Nordenskiöld (the first to make a complete crossing of the Northeast Passage), also contributed, publishing his snow photographs in France and Britain in the 1890s before dying of tuberculosis a few years later, at the age of only twenty-seven. Nineteen-year-old Bentley may not have been the first when he finally got it right and produced a successful photograph in January of 1885, but he was uniquely tenacious and persistent. Year after year he continued to take new photos in parallel with his life as a farmer, his actual job on the family farm he'd inherited. He was an amateur, in the original sense of the word: an *amatore* – one who loves. Self-taught, without any formal training, he nonetheless had a passionate approach to his subject. By the end of his life, in 1931, after almost half a century behind the camera and active to the end, he had amassed more than five thousand images that he felt were of such high quality that they should be preserved for posterity. Gradually, his fame grew and his pictures were published in well-known newspapers and even in scientific journals, such as *Nature* and *Scientific American*, with accompanying texts that he'd written himself.

Bentley was well aware that, for the general public, snow meant transformation and dreams. Words like 'beauty' and 'magic' are repeated throughout his writing. One of his contributions to *National Geographic* in 1923 was entitled 'The Magic Beauty of Snow and Dew'. He called the snowflakes his 'snow flowers'. He made reference to precious stones, jewels, diamonds. For him, this was no exaggeration. The snowflakes made Bentley a literal fortune through selling photos to clients, but he also generously donated them to institutions and private individuals. The pictures became collectors' items, which pleased him enough; getting rich from the photos had never been Bentley's objective. 'I wouldn't trade places with Henry Ford or John D. Rockefeller if I had their millions. I have my snowflakes!' This added to his status and contributed to the wave of popularization that snow underwent in the early decades of the twentieth century.

Crucial to Bentley's success was his collaboration with George Henry Perkins, a professor of natural history at University of Vermont. Together they wrote a paper that was to achieve great fame, arguing that no two snowflakes are ever exactly alike. It was neither a new nor particularly important idea (and it has been debated whether it's even provable), but it was given a new lease of life by being presented using the new technology of snow photography. Through the widespread media interest in Bentley's snow images, the pair's claim became a truism among the general public, and in so doing, literarily useful (e.g. Pamuk). This collaboration and their impactful concept contributed significantly to both Bentley's authority and his popularity. He was, after all, just an amateur photographer, self-taught on his family farm. But by collaborating with a scientist, he gained greater credibility, as well as status from being associated with Perkins's institution. At the time, the number of experts on snow was small, and there were few institutions anywhere in the world that were specializing in this area. So when it came to snowflakes, all eyes turned to Bentley, and he was soon not only publishing articles in scientific journals, but also given the honour of writing the article on snow in the *Encyclopaedia Britannica*.

Bentley's major work, *Snow Crystals*, came off the press just a couple of weeks before he was to die of pneumonia, in December of 1931. He was caught in a blizzard following another photographic excursion and the two-mile-long walk back home was too much for him. The book contained reproductions of some 2,500 of his images and was based on another collaboration, this time with physicist William J. Humphreys of America's National Weather Bureau, who'd written a book entitled *Physics of the Air* (1920) and had the expertise to recognize the value of Bentley's work. Humphreys had also been interested in clouds, fog and artificial rain, publishing his book *Rain Making and Other Weather*

Vagaries in 1926. In a minor study on the role of volcanic eruptions in climate change, published in 1913, he asserted that volcanoes had lowered the temperature at the earth's surface several times since the eighteenth century, and that it was reasonable to assume accordingly that volcanoes had been a central cause of climate change throughout the earth's history – and would continue to be so in perpetuity.

Wilson Bentley has become the iconic representative of the casting of snow as a snow *crystal*, especially in North America and therefore also in mass media. At the same time, his importance to the science of snow is relatively modest, as he occupies a space between two eras in the modern history of snow. First, he belongs to the long tradition of crystal-seekers, whom he joined late. Without a doubt, collecting and documenting individual snowflakes, cataloguing their shape and uniqueness, was still a viable project at the time. Bentley contributed an overview of the subject and a large collection of observations. But so did other people; what he was doing wasn't particularly original any more. The scientific contributions came from his collaborators, while Bentley's major role was to take the pictures and embody the role of his nicknames, 'the Snowflake Man' or 'Snowflake Bentley', as which he was a kind of celebrity.

Bentley was the last in this line of snow researchers who lacked both training and a laboratory, which was why it had been difficult to conduct scientific research on snowflakes. While physics and chemistry experiments could be conducted indoors, at temperatures above 0°C, and it was possible to create intense heat artificially, cold was a different story. It was only when modern refrigeration technology became widely available that it was possible to create the delicate snowflake within laboratory environments. This whole new creative phase – another new beginning in the history of snow – picked up speed at almost exactly the same time that Bentley was capturing his final images with his camera in the late autumn of 1931. The impact of the new indoor refrigeration was striking. The centuries-long research into the shape of snow crystals ceased almost immediately. Their photography came to an astonishingly fast halt, too – not completely, but nonetheless the focus now shifted to reproducing and visually recording experiments in the laboratory.

Meanwhile, an outdoor revolution was taking place. Like Ukichiro Nakaya in Japan, Swiss scientists working on snow soon concluded that what mattered was to approach the substance in a greater number of ways, using more sophisticated methods. One of their first laboratories, which was made entirely of snow, was set up in the village at Davos in the winter of 1935. The following year construction began on a research station that would use field instruments, this time at Weissfluhjoch, at an altitude of 2,536 metres. The site was close to a railway station and therefore easily reached from Davos village, which stands at an altitude of 1,500 metres. There was also a large number of avalanche-prone slopes in the immediate vicinity. One of the primary objectives was to deepen the understanding of avalanches, how they form and how they behave, though the measurements focused on all sorts of meteorological and physical geographic conditions. Research was later extended to include the mechanics of snow, snow transformation and the properties of snow under different weather conditions. The scientists also tested equipment and instruments, and carried out experiments, including avalanche tests.

What is characteristic of this period is how quickly the research repertoire widens. It becomes important to see the bigger picture. Research into snow – or at least some kind of rational speculation about it, which had been going on since Aristotle and, in a somewhat more systematic form, since Kepler and Descartes – is no longer just about finding, imaging and sorting unique snow crystals. That line of work, which had dominated for more than two hundred years, continues, particularly in Japan, but it now diminishes in importance – something that begins almost as soon as Bentley pulls the last frame out of his camera.

The shift in the way interest in snow manifests at this time is even more evident if we look at all the other research on the physics and chemistry of the earth that's now growing rapidly and finding its way out into general society. More and more learned societies appear whose publications sometimes have 'ice' and 'glaciology' in their names, but they also welcome research on snow. The International Commission on Snow (ICS), formed in the 1930s on the initiative of James E. Church, a professor of classical languages in Reno, Nevada, traces its roots back to the 1890s. Church, an active climber in the snow of the Sierra Nevada mountains, developed a global snow humanism ('I sought snow, and found peace and understanding among men'), viewing snow as both a critical water resource and as a path to peace and cooperation, because snow and glaciers are to be found in mountainous regions shared by more than one country. A *Zeitschrift für Gletscherkunde* ('Journal of Glaciology') was founded in 1906, but the last issue was published in 1944. After the war, the editor, Austrian Raimund von Klebelsberg, no longer had any international contacts and German-language journals were in chaos. A new journal, the *Journal of Glaciology*, begins publication in 1947, with a foreword in the first issue by the highly respected Swedish glaciologist Hans Ahlmann, now at the height of his career. It's becoming clear that snow is more present in the realms of public interest than ever before. Which also means that there are greater demands being placed on how snow is represented visually. Fascinating images are not enough. The investigation has to go deeper and, sooner or later, it has to be carried out by professional researchers asking more fundamental questions.

That's why we must now turn to Japan, where great things were happening. At almost exactly the same time that Bentley left the scene, along with his beloved camera, a laboratory was founded at a new university in the city of Sapporo, on the north Japan island of Hokkaido. The laboratory was led by a young Japanese physicist. Whenever it snowed, as it often would, he went outdoors with his assistants and collected flakes to put under the microscope. He noticed that the snowflakes formed similar basic patterns for a few minutes, then they began to mix with another structure, which was gradually replaced by a third. The crystals seemed to reflect small changes in the weather. No one could imagine what was going to happen next.

The young physicist was Ukichiro Nakaya, the man whose daughter's art I experienced in Oslo and whose snowflake photographs I've seen with my own eyes. After studying in Japan, during the 1920s Nakaya was accepted for postgraduate study at King's College London. He was talented, but after graduation it wasn't clear how he would continue his career. He was keen to take part in the international research community, but how would he go about such a thing? In London, he had been studying long-wave X-rays, but he and his family needed to make a living.

This, in a nutshell, was the background to Ukichiro Nakaya's decision to settle in Sapporo in 1930. At the tender age of thirty he was appointed Professor of Physics at Hokkaido University, and rumours of his talent soon gave rise to great expectations. The conditions weren't the best at the small and under-equipped science faculty, which had been created earlier the same year, and Nakaya had few colleagues in his department. However, he soon realized that the situation could enable him to forge his own destiny in this strange, but also lovely and mesmerizing environment, which he generally liked and thrived in.

But what would he focus on? Despite everything, there were still some advantages to his situation: he had enormous freedom to design his own research, and the university had promised him the resources to build a laboratory for cold-temperature experiments, which both he and many colleagues elsewhere at the university wanted. What's more, from his time studying in London, he had an extensive network of colleagues around the world. And, perhaps most importantly, he had plenty of snow. Northern Japan is one of the snowiest regions in the world. Low-pressure systems march in from the Pacific Ocean, the dense clouds having a high moisture content. As they pass over the mountains of Hokkaido, which at their highest stand at around two thousand metres, snow falls – often for days on end. Snow depths of several metres aren't uncommon and Hokkaido holds world records for both accumulated snow-depth per season and greatest snowfall in a single day.

After a couple of years of reflection, Nakaya decided to make a virtue of necessity, and thus snow became his object of study. Perhaps that wasn't so strange. Some of his work at King's College had involved analysing crystals, and snow stars are crystals, too. Also, Nakaya had photographed snow crystals in a laboratory during his first winter in Sapporo. Over the following three winters, he continued his work in a cabin called *Hakugin-so* ('the Silver Villa') halfway up the nearby 2,000-metre-high Mount Tokachi in the centre of Hokkaido, taking thousands of pictures of snow with his colleagues. He would then spend two winters in an even smaller cabin on the slopes of Mount Asari near Sapporo, in conditions that were downright inhospitable, making research very challenging. Nakaya and his assistants were plagued by the cold. He was at times so ill that a couple of colleagues had to take the reins. Worse still, the research results were disappointing. Nakaya later wrote, 'We could not as yet begin to explain the simplest and most primary problem, why snow crystals show such a complicated variation in form and structure.'

Nakaya had been approaching snow from a natural science point of view – collecting and recording snow as it looked in reality. Empirical outdoor snow. This might have been the right approach if the aim had been to understand larger geophysical phenomena requiring extensive observations. But it wasn't particularly helpful in addressing Nakaya's central question: what lay behind the variation in the internal structure of snow? In just a few years he took more than three thousand photos of snow, almost as many as Bentley took in a lifetime. Using these images, he was able to classify the snowflakes, but he still didn't understand why they looked the way they did.

When progress failed to materialize, a new vision began to emerge. What if it were possible to create *new* snowflakes? Through the use of a new, experimental method, Nakaya began to understand what forces determined the shape and structure of snow stars. He then decided to work more like a bacteriologist and 'cultivate' the objects of his study himself. But this was easier said than done. Naturally formed snowflakes float freely in the atmosphere as they evolve, and their slow twirling towards the ground gives them plenty of time to grow, sometimes to considerable size. Of course, it was impossible to build a vertical freezing facility a couple of kilometres above the ground to monitor the development of snowflakes, so Nakaya had to find another way to produce the crystals he needed.

By this time, it was clear to researchers that the crystallization process is made much easier if there is a crystallization 'nucleus' – a grain of dust or dirt that can start things off. This was something that Johan Carl Wilcke had been working on in Stockholm back in the eighteenth century, but without having access to the atomic theory that would emerge in the following century and became profoundly enhanced in the early twentieth century. The theoretical context available to Nakaya was infinitely better. In practical terms, though, it was difficult to make individual snow crystals. He tried to create such a crystal on fine threads, or filaments. In this way, the growth of the individual snowflake could be observed in detail and the artificial 'atmospheric' conditions controlled by Nakaya himself. It didn't quite work out the way he wanted. What he got was a large number of tiny ice crystals that covered the entire thread with a layer of frost, preventing the detailed observation he was hoping for. But Nakaya didn't give up. He patiently made his way forward, trying a variety of filament materials – wool, silk, cotton, even cobwebs. But all he got were these frost-like clusters of crystals.

Other studies in his laboratory were more successful. He constructed special cylinders to determine the drift speed of different crystals; this was found to vary widely, by a factor of ten. He studied the electrical properties of snowflakes. He also identified previously unknown types of snowflakes in natural snow. The volume of observations allowed Nakaya and his team to correct many of the simplifications in previous research. Gradually, a conclusion emerged that while snowflakes reflected the state of the atmosphere in which they were formed, the same sky, at the same time, could produce greater variation in snowflake structure than had previously been thought possible. The snowflake continued to frustrate researchers.

In 1936, the breakthrough came – in the form of a rabbit hair. The rabbit hair had a microscopic roughness to its surface which offered specific crystallization points, allowing Nakaya to avoid the general frost formation that had occurred with the other materials. This is how Nakaya was able to grow snow crystals individually, under conditions similar to those when natural snow forms inside clouds high up in the atmosphere, to which he had no experimental access. It was a major breakthrough. Nakaya was able to mimic the properties of clouds in his laboratory, creating the world's first synthetic snowflake. By then, it had been five years since he and his team had started their snow research.

One of the things responsible for Nakaya's success was the new low-temperature laboratory in Hokkaido, which opened in the year they had the breakthrough, enabling Nakaya to do more detailed experiments with different kinds of materials. In this space, the temperature could be brought down to -50° Celsius. It may have been cold in the small research huts up in the mountains, but it was nothing compared to what was possible in the low-temperature laboratory. Work was regularly carried out at -35°C, sometimes at -45°C. The researchers had to sit at their instruments wearing flight suits. At these low temperatures, the conditions for making artificial snow were optimal, as long as you used rabbit hair! Once Nakaya understood what needed to be done, he was able to produce new crystals more quickly and conduct his experiments. By varying the temperature, humidity and other essential conditions for these artificial snow crystals, Nakaya created his own snow taxonomy that was both more accurate and more complete than anything that had come before. His hope was that it would help meteorologists determine the atmospheric conditions that shaped the snowflakes they found on the ground.

Nakaya shared his ambitions with a few select colleagues around the world who, like him, were paving the way for scientific discoveries about this eternal element – snow – which could at last be properly studied as a crystalline substance. Previously, snow hadn't attracted much scientific interest, and there was a general belief that its properties were similar to those of the soil and that there was little point in studying snow as a distinct category separate from water and ice – snowflakes were fascinating and beautiful, but they would inevitably become one or the other of those as soon as they landed on the ground. What, they wondered, was the point of studying just snow? Even snowflakes hadn't attracted much attention. In the book that he was perpetually working on about his snow research, Nakaya pointed out that since the efforts of Wilson Bentley and Antoni Dobrowolski a couple of decades earlier, hardly a single significant paper had been published. The research Nakaya and his colleagues were carrying out around the world in the 1930s soon proved that the earlier paucity was no cause for pessimism. In fact, snow science was entering a whole new era.

Nakaya was to be part of this new era. However, the war years meant a significant hiatus for him and his snow research. During that time he, like his European and American colleagues, was conscripted into research for the war. When Japan entered the Second World War with its attack on Pearl Harbor in December 1941, Nakaya was put to work on atmospheric icing and the artificial dissipation of fog that might help the war effort, while finishing his tome on snow in what little spare time he could find.

The preface of the finished book is dated 1944. While it was in production, the printer was bombed and the copper plates carrying all the text were lost in the flames – a terrible tragedy. Nakaya gave up hope of ever seeing his research published, but in the spring of 1949 he was unexpectedly asked by the American Academy of Arts and Sciences to write a new version of the book for an English-speaking audience. After the war, there was extensive reconciliation work between the United States and Japan, which now extended to Nakaya's snow research. The Academy's funding enabled Harvard University Press to publish his book in 1954, by which time

Nakaya had already completed an extended fellowship in the United States. The cover features a photograph of a six-pointed snow star.

Snow Crystals: Natural and Artificial is a remarkable, five-hundred-page tour de force, a comprehensive review of the vast world of both snow as a natural element and its man-made counterpart. Observations of snow, the classification of snow, the physical properties of snow, the artificial production of snow, and the investigation and comparison of different kinds of snow are all laid out in a text supported by finely drawn diagrams and some of Nakaya's most breathtaking images from the Hokkaido low-temperature laboratory. You get to see the author himself in his flight suit, hunched over his microscope. At the same time, it's a cosmopolitan book that uses the universal and timeless language of snow to mend the bridges between cultures and societies which at that time had been broken by years of war, hostility and racial prejudice. Nakaya begins the book not only by thanking his Japanese assistants and colleagues, but also by summarizing the many international precedents in a historical-research overview.

The book was vigorously promoted with advertisements in major scientific journals, emphasizing that it was written not only for specialists: 'many laymen will appreciate this authoritative and beautifully illustrated book'. Reviewers praised 'the poignant beauty and astonishing variety of snow crystals'. Yet the purpose of the images was neither aesthetic nor decorative. They provided the documentation Nakaya needed to establish his new scientific taxonomy of snow types – a superordinate 'snow order'. This has been likened to a snow periodic table. Many pieces of the puzzle were already in place, and Nakaya had found several of the missing ones. Yet snow was not a 'system'. There was something more to snow, which was yet to be found.

Nakaya's main contribution to his field was undoubtedly the transfer of snow crystal research into the laboratory, which was also the focus of his book. One of its reviewers commented that one of the most satisfying achievements for Nakaya and his group must have been the 'artificial production of crystal types that had never been found in natural snow'. Kenneth Libbrecht at Caltech is full of admiration for Nakaya in his own book *Snow Crystals*, especially when it comes to the so-called Nakaya Diagram, a striking picture of how a snowflake's growth conditions control its shape. Libbrecht points out that the diagram has been likened to 'a Rosetta Stone for snowflakes'; a generalized code to determine the meaning of snowflake 'hieroglyphics', just as the Rosetta Stone had done for the ancient Egyptian language in the ancient world. Nakaya is often quoted as referring to the snowflakes as 'letters from heaven'. The word 'hieroglyphics' puts more emphasis on the need for interpretation, but the idea is the same: that the structure of the snowflake carries information about the circumstances of the exact time and place in the atmosphere that it was formed.

This was a beautiful and poetic idea that could be understood by anyone, even the vast majority of people not well-versed in the physics of snow formation. Before the war, Nakaya's ideas had not travelled very far outside the narrowest of specialist circles. After the war, thanks to the publication of his *Snow Crystals* book in the United States, they were widely disseminated. Back home in Japan, Nakaya was already a highly regarded educator and writer, combining information about the physical properties of snow with vivid depictions of its beauty and magic. His books for a wide audience had titles such as 'Winter Blossom', 'The World of Frost Trees' and 'Spring Grass'. He depicted snow visually in photo books and wrote many books for children, such as 'The Cold Country' and 'Snow in Hokkaido'. He appeared in short scientific films, such as 'Snow Crystals' and 'Frost Flowers', and also co-founded a film company to popularize the natural sciences.

Snow had an ancient tradition in Japan, with roots dating back to the nineteenth century and even earlier. Yet Nakaya can be seen as a Japanese example of the same modern winter nationalism we have seen in the Nordic countries. Nakaya was in many ways a Japanese Fridtjof Nansen – versatile, artistic, popular, an aesthete and an outdoorsman, a naturalist in the laboratory, but also a humanist with a conscience and a desire for peace and international co-operation. An important difference is that Nakaya appeared at a time when another version of snow was taking shape, in the guise of the Cold War and in a world where snow took on new meanings. Post-war snow science required delicate combinations of realpolitik and pacifism, but also empathy. Qualities that Nakaya fully embodied.

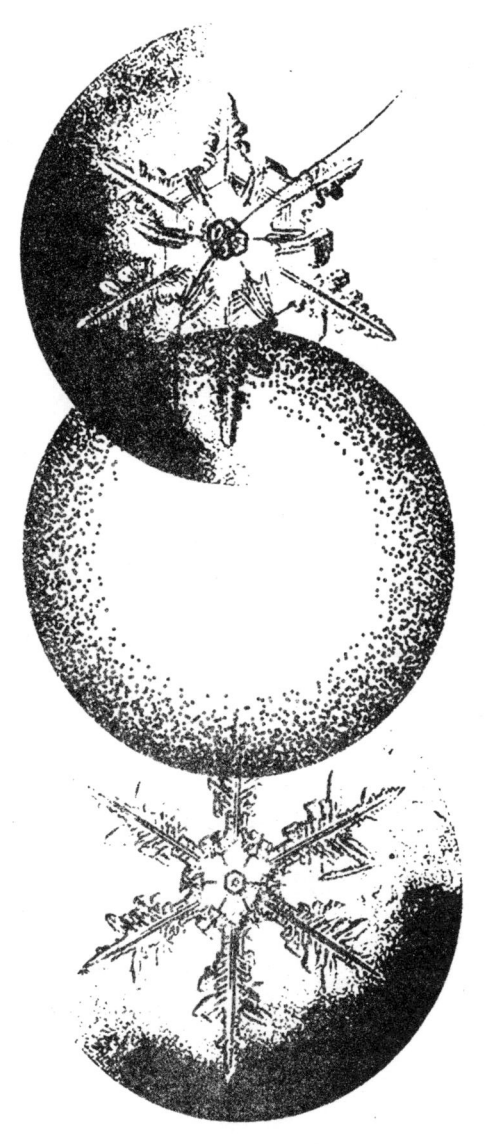

Strategic Snow

*

MY TERRY-CLOTH SHOWER TOWEL HAS A PICTURE of an igloo and small children that we're still referring to at the time as 'Eskimos'. I use it when I'm showering and taking a sauna in the low, dark-red-brick building where we have our PE classes. Grey and white tiles. The Eskimo children's traditional dress is a liberating red and orange. I've seen Greenland on the globe – know that it's big and close to the North Pole. There are polar research stations up there. And to think that there is even more snow there than where I live! I feel kinship with the Eskimos. We're the world's snow children.

I watch Disney films about snow on TV. We learn about Nanook in school, in American producer Robert Flaherty's feature documentary from 1922, *Nanook of the North: Son of the Cold*. The art of cinema itself seems to be born in snow. The Arctic and Greenland feel close. Flights to America take the 'North Pole Route', Dad says. We know that there are submarines under the ice – a creepy thought that I push away.

At that point, I can't tie the various information strands together into a context. But there is no mistaking that the contemporary significance of snow is strong enough to trickle all the way down to my own primary school in Åsele, in the centre of Sweden's Västerbotten province. On a burlap board on the wall in my classroom there are many small snow stars hanging, fixed there with pins, one for each student in the class. You get a woven ribbon to add to your pin for every mile travelled on skis. I have to count every mile I ski, both in class and outside it. It makes for a thick bunch of ribbons.

'Snow Star' is a campaign by the Swedish Ski and Outdoor Association, whose membership is growing rapidly at this point in time. I use the word 'snow star' daily. In class we're taught that snowflakes are six-pointed and extraordinary and that no two snowflakes are ever alike. Or almost never . . . that part seems unclear.

What we don't learn much about is what an incredible substance snow is. A transitional substance, water at its heart, but also closely related to ice. Together, snow and ice compose the fragile cryosphere, the frozen parts of the earth. But I don't know that then. Nobody seems to know that. No teacher utters the word 'cryosphere'. In the geography textbook we learn that 'climate' is about how terribly cold it is in Verkhoyansk in Siberia, and how much it rains in Cherrapunji in India. Metres and metres of rain. That's the way it is in India. The climate does not change. Same with the snow in Greenland. We are still in the silence of the world before the great warming. But there had been whispers about it; about the snow that seemed so eternal. Could it ever melt away, never to be seen again? Could the ice disappear?

In 1932, a Danish geographer calculated what would happen if the Greenland Ice Sheet melted. The world's oceans would rise by eight metres; all of Copenhagen would be under water. 'Only the area around the Church of Our Lady and the

University of Copenhagen would remain, as an insignificant island.' The geographer, whose name was Einar Storgaard, had based his work on the German geologist Alfred Wegener's expedition to the highest point of the Greenland Ice Sheet, *Eismitte*, in 1930–31. The versatile Alfred Wegener was a serious researcher who had already been to Greenland three times and was a pioneer of ice analysis – and the same man who in the 1910s put forward the theory of continental drift, which is now widely accepted. The expedition was delayed by unusually thick and difficult sea ice and bad weather, and only some of the party could be sent on to the highest point. The propeller-driven sleds, *Propellerschlitten*, which had initially been highly lauded, didn't work and had to be abandoned. Three participants stayed over the winter with limited supplies; Wegener and a Greenlander, Rasmus Villumsen, meanwhile, had to return to winter at the fjord, a journey of four hundred kilometres. It was already November, the conditions were intolerable, and they both died during the return journey. Wegener died first and was carefully buried in the snow by Villumsen, in a stitched reindeer skin marked by a cross of skis. Villumsen's body was never recovered.

The three who stayed, Johannes Georgi, Fritz Loewe and Ernst Sorge, knew nothing of their colleagues' tragedy as they spent the winter at what was named *Station Eismitte* (Mid-Ice Station), which was actually no more than an unheated cave in the snow. There were dedicated spaces for a barometer and a seismograph – the latter for studying the thickness of the ice sheet – and room for the three men's daily lives. Most of the time they spent in reindeer-skin sleeping bags as the temperatures crept down to -50°C or even lower. On top of this igloo of science, the three erected a tower several metres high used to launch observation balloons. Their diaries and accounts of their year at the station speak of anxiety and monotony, of amputated toes, but also of deep connection, meandering conversations about the riddles of life, Schopenhauer's philosophy, war and peace, how they missed everyday life. Or how they just wanted to stay alive.

The research programme focused on snow accumulation and snow-melt, but had to be reduced in scope due to their circumstances. The three had dug a shaft more than fifteen metres deep in the snow to inform their study of snow stratification. The depth corresponded with twenty years of packed snow, going back to 1911. To get up and down a shaft of that depth, they built a slope at about 50 degrees, though the last five metres into the darkness were vertical and so narrow that only one person could fit. Working in the shaft was Sorge's task, and it was a gruelling and risky exercise. But also quite fruitful. Sorge found that in this extreme environment, which experienced low temperatures all year round, the snow wasn't as clearly layered as it was, for example, in the Alps, where snow season and melting season alternated each year. Sorge weighed blocks of snow that he cut out from the site vertically, and found that there was still a clear difference between the seasons, if you took the density of the snow into account. Summer snow was lighter, winter snow heavier, and accordingly packed harder. Snow wasn't a clear-cut and homogeneous substance, influenced only by the force of gravity, even in an environment without melting. It carried signatures of the condition of the atmosphere. The results were in line with what was then the current knowledge about snowflakes and snow, but on a much, much vaster timescale – not just an instantaneous measurement, but something spanning a multi-year period. From this, thought-provoking conclusions emerged: snow and ice environments might not just have things to tell us about small, fast changes in the weather – they may also tell us about big, long-term changes.

However, the seismographic part of the programme was a failure. A support team (one that effectively became a rescue expedition, in the summer of 1931) brought sleds full of explosives and a large number of blasts were carried out, sending snow swirling hundreds of metres into the air. But the seismograph didn't work as intended and the return waves coming from the bedrock weren't clear enough to lead to firm conclusions. That didn't stop Sorge from making his own pronouncements about ice thickness, which turned out to be fairly accurate. He estimated an ice thickness of up to 2,700 metres; in reality, it was just over 3,000 metres. Taking the two sets of information together – the link between snow stratification and atmospheric changes, and the enormous thickness of the ice – it became clear that this massive sheet of snow and ice must be very old. If so, could it also contain a wealth of information about the climate of the past – and how it changed?

The expedition had ended in tragedy and the research programme was largely a failure. But still, the results collected at *Eismitte* opened up new possibilities for under-standing atmospheric change over very long periods of time. The seemingly boring and changeless snow in one of the most unchanging places on earth turned out to carry within it information that might perhaps give insight into huge changes – across the globe and over centuries and millennia.

What they were faced with, in other words, wasn't just an ice machine of global proportions, but also a time machine that potentially could tell the story of times gone by in a way that few, if any, other materials could.

Henri Bader, the Swiss physicist, later dubbed the discovery from the snow-shaft site at *Eismitte* 'Sorge's Law of Snow Densification on High Polar Glaciers'. It was a machine that never stopped. Even in the time the three scientists had spent on the ice that winter, the snowpack had grown by three metres, so that by the time the rescue expedition reached them, the station itself was completely invisible except for the top of the balloon tower sticking out of the snow. Sorge's observations might seem trivial on the surface, but in fact their implications were far-reaching. The snow-and-ice-cover wasn't some inactive mass. Its stratigraphy, or layering, was full of every possible story about the past. Prior to this, glacier research had been mostly concerned with the consistent differences in the accumulation of snow, from year to year, and the corresponding melting. But what if these variations were preserved in a way that made the changes measurable?

What Sorge seemed to have created was 'legible' snow, a snow palimpsest of nature's own, with writing superimposed upon writing, year after year, perhaps continuing all the way down through the verticality of the ice? Glaciology suddenly opened up completely new horizons. The interwar period was warm in the Nordic countries and the North Atlantic region – much warmer than the previous centuries. That sparked the imagination. Did large-scale melting in recent years reflect warming caused by industrialized society? The idea came up as part of the debate around 'climate improvement', as it was sometimes called, although few people took it seriously. Could glaciology be a way of following more than what was happening right now? Might snow and ice be windows into the deeper recesses of time? The leading snow experts were cautious. Were people really melting the world's snow, and the world's ice? Unlikely . . . or was it?

Among those who built directly upon Sorge's work, and who saw the potential of the findings at *Eismitte*, was of course Einar Storgaard, when he fully extrapolated the facts and asked what would happen if the entire Greenland Ice Sheet melted. Another Dane who also saw the light in Sorge's tunnel was Willi Dansgaard, the physicist who'd wanted to know why the Norse settlers abandoned Greenland. It was his generation that, from the 1950s into subsequent decades, began to fully reap the rewards of the Wegener expedition's findings. Dansgaard followed the logical idea that the stratification in the snow was propagated through the firn and continued further down into the glacier ice. If you wanted to access the deep time record, you had to drill deep into the ice, where it would then be possible to trace changes in atmospheric conditions using different isotopes, a method he had developed.

Others beat him to that drilling, though. The first such expedition in Antarctica was led by the Swede Valter Schytt, one of Hans Ahlmann's students. The Norwegian–Swedish–British expedition, which went to Antarctica in 1949–52, went down 100 metres into the ice, which according to Schytt's calculations corresponded to seventy years. An American project on the Taku Glacier in Alaska attained the same depth. Expéditions Polaires Françaises (EPF) also drilled in Greenland, down to 150 metres. These early ice drilling projects focused on the borehole itself, which was used to access various aspects of the glacier and the older ice that it contained. Also, the drills available were unable to preserve the long ice 'cores' they removed, which in any case partially melted during the drilling process itself, and there was no continuous 'cold chain' that could have preserved them, so they were examined on the spot, under primitive conditions.

Dansgaard realized that the historical information he sought could only be obtained if the core samples could be better cared for and transported to well-equipped laboratories for analysis. By the end of the 1950s, these conditions were realized through a series of both intentional and somewhat random circumstances. They had learned to drill in a way that preserved the ice core itself, like a cylinder of frozen time. That was the engineering component. They had also learned to interpret and understand the isotopes and other atmospheric data enclosed in the ice core. That was knowledge that physicists, like Dansgaard, and chemists had mastered. And they had learned how to connect the drill sites located in the snow and ice fields with laboratories in Copenhagen, Paris and elsewhere, using aeroplanes, refrigeration and insulating chemicals. This was the kind of logistical support that the Cold War provided, in the form of expertise and resources, which also led to opportunities for collaboration between the practitioners of the new techniques in snow and ice – physicists and engineers from Denmark, Switzerland, France and the United States, all working together in some of the most inaccessible places on the planet.

In 1957, scientists from the Snow, Ice and Permafrost

Research Establishment (SIPRE), with Henri Bader as director, managed to drill more than four hundred metres into the ice at Site 2, a US military base in north-west Greenland. Now their focus had shifted to what would become their new concept of snow as a material that, once it's been stored within the ice, can tell us about our planet's conditions in the past – or, as the project's mission statement put it, about the 'construction and preservation of under-snow structures – in a nearly continuous profile'. Great care was now taken to preserve the drill core itself, which was intact with clearly legible layers for summer and winter. One of the most recognizable images of Dansgaard shows him in a cold-room with an ice core in his arms, holding it like an infant.

Glaciology, as practised by geographers and geologists aiming to study the distribution and horizontal movements of glaciers, had undergone a vertical turn. A ninety-degree shift into the great depths of time, furthering understanding of the world's history and, in a sense, its future, at a point when this historical data could begin to inform the climate models that were also starting to be developed around the same time.

The last thing Ernst Sorge did before departing the station was to erect a seven-metre extension to their balloon tower, so that the site could be seen if a new expedition wanted to come there to continue their promising work. His own prediction in 1932 was that such a thing would happen within the following ten years at most. Of course, he had no way of knowing what would happen later in the 1930s, not least in his own country, Germany. In fact, it would take a decade longer than that, a global war and the subsequent

Cold War, before the station was reactivated. This time it was under American leadership, and with completely new ideas about what snow could be used for and why knowledge of snow was strategically important.

We'll come back shortly to why this was. But first, let us take a quick jump forward to Ukichiro Nakaya, whom we've followed so closely, in the mid-1950s. What was he doing now? The snowflake master is still around – and he is right at the centre of the action.

As the year changes from 1956 to 1957, Ukichiro Nakaya and a research team find themselves high up on the Mauna Loa volcano in Hawaii, where the American National Weather Bureau has a station at three thousand metres and where their work is based. Their mission is to study the formation of snow crystals in an environment with unusually clear, clean air. By now, Nakaya's research has been well established; he may be the most respected name in all of snow research, having published his flagship book in 1954. The Hawaii study is just one in a long line of enquiries Nakaya has undertaken since the early 1930s, and it wouldn't be worth singling out if it weren't for one thing: its location. Mauna Loa is the mountain where, in the year after Nakaya's stay, scientist Charles David Keeling would begin his later famous measurements of carbon dioxide levels in the atmosphere. Measurements that gave the world the saw-toothed, steadily rising curve that would soon become an iconic representation of our world's culture of fossil-fuel dependency.

This is more than a coincidence. Admittedly, the two

research methods are completely different and deal with different things – the formation of snowflakes and the measurement of carbon dioxide. They represent completely different branches of chemistry and physics. But they are also branches of another tree – that of anthropogenic climate change. They show how different knowledge traditions can reach out to each other. Without having had much contact with each other before, around the middle of the twentieth century they form a new, collective understanding of how a change in the functioning of the earth is being caused by human activity.

And they do so at the same point on earth. When Nakaya is on Mauna Loa in the winter of 1957, the International Geophysical Year (IGY) 1957–58 is just about to start, and it is as part of the IGY that Keeling launches his measuring instrument. The IGY involves thousands of scientists from more than sixty countries. This happens in the middle of the Cold War between the superpowers, which was also to mark a great era of snow research. The world gets Sputnik, which paves the way for satellites that will eventually map snow and ice from above. The world gets more polar research stations – like the one at *Eismitte*, where Ernst Sorge lays a foundation with sawed-out snow blocks and where kilometre-long drill cores will soon be extracted from the ice; or those built in Antarctica, which become bases for research that is the stuff of downright science fiction.

It was no coincidence that I had a so-called Eskimo on my towel up there in Åsele. A new purpose of snow is emerging: snow used for strategic purposes, something widely promoted in polar and space propaganda. But the same snow will also form the basis for the biggest strategic question of all – that of humanity's relationship with our climate and, in a broader sense, with the future of the entire planet, on which we all depend. This, too, is a strategic challenge, though it has not yet become strategic in the conventional sense. It is an insight in the making.

In 1957, after the project in Hawaii, Nakaya was in Greenland conducting research on snow and ice as part of the IGY. He continued to visit Greenland for the next two years on behalf of SIPRE, Henri Bader's institute. There he participated in studies of the properties of ice and snow at depths of almost a thousand metres, including the viscosity and elasticity of the ice. Studying what he assumed to be a pristine environment, Nakaya was surprised to find 'polluted atmospheres', with particles from industrial society. He had also found these in Hawaii. Nakaya realized what Keeling's Mauna Loa research was all about. It would soon become clear that carbon dioxide was abundant even in the most remote places on earth, and that the amount was systematically increasing. Which must mean that the earth was indeed becoming a greenhouse. Nakaya was one of the scientists who recognized the political relevance of the greenhouse effect early on:

Because of the global increase in atmospheric temperature since the beginning of the Twentieth Century, glaciers in many parts of the world are shrinking. The cause for this is an increase of CO_2 due to the automobile-dominated society and the cutting down of forests. Warming of climate will melt the ice in the Antarctic and Greenland, leading to a sea-level rise, and lowlands all over the world will be in danger of being submerged.

This was exactly what Einar Storgaard had predicted a quarter of a century earlier.

From the mid-1950s onwards, an increasing number of scientists began to argue that it was the increase in carbon dioxide that was leading to a warmer climate. The once somewhat lonely voice of Guy Stewart Callendar, the early advocate of carbon dioxide as the driver of warming, began to get more support. Nakaya's statement was one of the first expressing concern about what a human-caused warmer climate might mean, especially for a nation such as Japan, which had large population centres in coastal cities.

In 1959, Nakaya joined a group of American and Canadian scientists who had set up their experiments on a huge iceberg near the North Pole. The iceberg, known as T-3, was discovered in 1947 and had been an outpost for both military and civilian research since the early 1950s. Here, Nakaya would study the polar winds, ice and ocean currents that had an impact on the melting.

Ukichiro Nakaya died in Tokyo in 1962, just as his daughter Fujiko gave her first solo exhibition of paintings at the Tokyo Gallery. She had graduated with a degree in art from Northwestern University in 1957 and continued her education in painting in Paris and Madrid until 1959. Her early artistic career must have appealed to her father. He himself had become an accomplished practitioner of classical

Japanese ink painting, *sumi-e*, and father and daughter had even organized a joint exhibition of their ink paintings.

After the death of her father, Fujiko turned to water vapour or fog as her primary medium. She got off to a flying start with her famous fog installation in the 1970 Osaka World's Fair's Pepsi Pavilion, which also included works by Andy Warhol and Swedish-born engineer, entrepreneur and artist Billy Klüver. Fujiko's installations built on her father's research into atmospheric snow, for which water vapour is the precursor and most basic condition. One might even say that she's looking behind her father's crystals, to a pre-crystalline state where everything is still open. Where the final crystal hasn't yet taken shape and where art can still work with an openness both backwards in time and forwards, with expectation and hope as important and deeply existential elements. In this free application of her father's work, Fujiko Nakaya has also been able to give space to his concern about humanity and its destiny.

Ukichiro Nakaya's work embodied a difficult, but also productive, phase in the history of snow. He was part of the transformation of snow, making it more significant and more important than ever. At the beginning of his career, it was still mostly a snowflake, an aesthetic detail, a strange, lingering enigma in a world increasingly characterized by determinism and predictability. Captivating and dreamy, but ephemeral.

Thirty years later, the properties of snow were of the utmost importance in understanding the things about the planet that humans had begun to fundamentally alter. There was still magic, but it was dark magic made by humanity itself. Everything had changed in an astonishingly short time. And the word that was increasingly used when describing these kinds of changes was 'environment'. New laboratories were springing up everywhere, addressing different aspects of this concept in new studies that all pointed to this fundamental relationship between humans and change, and snow was sometimes present in their repertoire.

These were new beginnings. When looking at this period as a historian, it's striking how much was going on at the same time – all these various paths leading towards the new insight, without any prior planning or coordination. The logic and ambitions varied, depending on where and by whom the work was done, but there were patterns emerging in the work taking place. The very word 'strategic' turns out to have different meanings. Immediate military. Long-term planetary. There were national styles and approaches in the growing interest in snow. The 'budget balance' of glaciers – the building up and melting – took hold in the Scandinavian countries that had clear seasonal changes. In the Soviet Union, there was a strong focus on permafrost and sea ice due to the economic and geopolitical significance it held. Work on snow was also conducted by universities, the powerful Russian Academy of Sciences and, not least, the archipelago of Arctic-focused Soviet research institutes that held a globally dominant position in the interwar period.

Crystal research more generally was a high priority, in part for military reasons. The Normandy invasion of June 1944 had been planned with the help of Britain's leading X-ray and crystal scientists, including Cambridge chemist John Desmond Bernal, who attempted to assess how the sands of the French Atlantic coast might affect the military beach landing there. Bernal also advised on the British Habakkuk project, in which the construction of 500m-long floating airstrips, made of ice, mud and wood – a kind of simple aircraft carrier – were planned to fight German U-boats in remote (and icy) parts of the North Atlantic. It never came to fruition, but through his involvement in the project, Bernal was also able to ensure that his colleague and friend, Austrian-born Max Perutz, later a Nobel laureate and, like Bernal, a radical, was soon released from internment in Canada. Perutz, who had been analysing glacier movements, instead got to carry out experiments on ice crystals in the basement freezer of a London butcher.

Bernal didn't think strategically only about crystals. He thought strategically about science as a socio-political project. He was one of the first people to consider how humankind might colonize space. Most notably, in 1939 he published the first true bible of science planning, *The Social Function of Science*. Bernal was convinced that science could help win the new 'Great War' and wars to come, despite, or perhaps because of, the fact that he had long been a devout pacifist, deeply suspicious of the actual capabilities of the military.

When it comes to snow and cold, it was the Americans and Soviets who took the lead, because their strategic interests in these subjects were the greatest. In the far north, the United States and the Soviet Union, and their respective allies, had started preparing for a third world war, which would take place in the Arctic, because the superpowers were closest to each other there. That's why the Americans stayed in Greenland after the war, continuing to build military bases and installations, including missile silos in the ice, and why the Soviets built new bases and invested heavily in Arctic research.

American scientists were fully aware that their own knowledge of snow and ice was underdeveloped compared to that of both the Soviet Union and other European countries, and the US military was equally aware of its lack of glaciological preparedness. We could have been facing a 'scientific Pearl Harbor', as a 1958 report described it, referencing the possibility of a devastating defeat resulting from a lack of knowledge about conditions on the Arctic battlefield. A few years earlier, a survey had shown that the number of people trained in glaciology at the hundreds of American universities was a scant forty. By the 1950s, the Americans had ten thousand men in Greenland. If conflict broke out, that number would multiply many times over. Squadrons of ships and planes would sweep in from all directions, with soldiers having to orient themselves without any experience in a terrain with no towns and no vegetation – completely covered in snow. This was the Arctic 'environment' and they weren't prepared for it. The needs were enormous.

Simultaneously, the requirements were different from anything humanity had ever needed before from snow. The snow and ice of hyper-advanced, modern war wasn't the fluffy snow that had floated down over Hokkaido or stuck to Wilson Bentley's photographs. Nor was it the snow of Elsa Beskow's children's books or the silent snow found in the landscapes of Nordic national painting. In the US military's arctic, snow was at once a lurking threat, a deceptive whiteout – and a prosaic building material for heavy infrastructure and logistics. Military documents from the late 1940s up through the 1960s reiterate how strategically imperative it is to have basic knowledge of how snow and ice work in order for these elements to support war. This was an engineering task. Engineers would build roads, airstrips, military bases, shelters, missile sites, fuel depots, combat centres and other facilities, all on the snow, ice and tundra. Some of them within the ice itself. Snow was the military polar world's equivalent of brick and concrete.

To achieve these ends, they needed to know more about how snow actually behaves. And, oddly enough, given all the research into snow that had been underway for such a long time, there was very little applicable knowledge. Snow had been defined in many different ways, but most of that knowledge resided outside the US. Ironically, one resource available domestically was the knowledge of the indigenous peoples; the Inuit and other Arctic peoples on the American continent had always built homes out of snow. They had also, for generations, managed to orient themselves over vast distances with a minimum of landmarks.

Some learned from the Arctic peoples. Perhaps the most famous of these was the Canadian anthropologist Vilhjalmur Stefansson. In the 1920s he had published the visionary books *The Friendly Arctic* and *The Northward Course of Empire*, emphasizing the superior adaptation of indigenous peoples to living conditions in the Arctic. Later, as a professor, he taught Inuit snow-building to students at Dartmouth College in Hanover, New Hampshire. But Stefansson was the exception, and he was eventually declared *persona non grata* by the FBI, which saw him as a security risk because of his alleged contacts with left-wing organizations during the worst of the Cold War witch-hunts.

A solution closer at hand for the US military was to look

to friendly countries that had developed basic snow knowledge, such as Japan, France and Sweden. That's how Hans Ahlmann's voice got amplified all the way to the Pentagon. Ahlmann's expertise on glaciers and climate change was all the rage in the US military. Through the efforts of his compatriot in the United States, meteorologist Carl-Gustaf Rossby, in 1947 Ahlmann received an invitation to visit the Pentagon and tell them what he knew about these matters. This knowledge had an impact on the design of airfields and general thinking about military installations in the strategically important Arctic. It explains the presence of Nakaya at US institutes and bases. But there was one country, above all, that had cultivated the kind of expertise on snow that the US urgently wanted more of. Knowledge that boosted both practical applications and strategic thinking. And knowledge that worked from the ground up, by examining the physics of snow. That country was Switzerland.

Down-to-Earth Utopia

*

HALF A LIFETIME LATER, I'M SITTING ON A BUS in New Hampshire. It's already a new century. On my way from a college where I gave a lecture, down to Boston to catch a flight home, it starts to snow. It snows more. Anxiety spreads throughout the bus. Will we make it or get stuck in the snow? I wonder if I'll make my flight to Europe. We reach Boston, but the airport is closed due to snow.

At breakfast in my hotel the next day, I read an article in the *Boston Globe* about a book written by the pianist, composer and lecturer Allen Shawn. The book is called *Wish I Could Be There: Notes from a Phobic Life*. I immediately push out into the deep snow, over to Harvard Book Store on the other side of the Charles River. You know that feeling when you just have to have a book? Like I've been waiting for just that book my whole life. And it's there!

Shawn describes how his anxiety burst into being when he was a young person. Preventing him from flying, riding in elevators, walking through windowless hallways. Eventually it controls his life, like a drug. Hence the title, *Wish I Could Be There* – because he almost never shows up, is never really present. He gets all kinds of care from the best doctors and psychologists available. Nothing helps. So ultimately he has to live his life with his phobia.

I turn around, too. On my way to passenger ferries, ticket in hand. On my way into the subway or to take buses. In London, I take taxis; ideally I walk. I can't attend mass at Notre-Dame. Too many people. I avoid going to concerts. Have to have an aisle seat at the cinema. But I never retreat from snow. With the help of snow, I can escape.

In the snow, I'm free.

The history of snow includes some specific clichés that have been around for a long time. For example, the idea that two snowflakes, though they always look fairly similar, can never be *exactly* the same. That's a far-fetched claim – but a beautiful one: the thought that what appears to be an infinite monotony of whiteness and symmetry is really an infinite diversity of expression. But is that true? What does it mean to be 'the same'? We know better now how to approach this ultimately controversial claim. I promise I will reveal why before we come to the end of this book.

Another, similar, controversy revolves around how many words there are for snow among the Inuit people in Greenland. The cliché is that the Inuit have a hundred words for snow (perhaps even two hundred ...). Its equally clichéd opposite contends that this isn't true, but in fact an urban legend. The answer to this controversy I will reveal right now: it's not an urban legend. It is quite true that the Inuit have many words for snow. However, exactly how many depends on how you define 'word', because many of the words for snow come in a number of variations and derivations, so it can be difficult to determine which are actually additional words with distinct meanings.

In each case, there is an origin to the cliché. I'll be returning to the one about the distinctiveness of snowflakes. Snow words among Arctic peoples (I'm expanding the geographical region) have been the subject of plenty of investigation precisely because the subject has long been controversial. A large number of snow words are included among what are sometimes called 'white lies' about the Inuit – in fact, not white lies, but *whites'* lies. 'Whites' because the anthropologists and other experts who consider themselves qualified to comment on the matter are usually white. The number of snow words in the Inuit language Inuktitut is often given as fifty-two, but many anthropologists and linguists have called this number into question, citing studies that have given much lower numbers, as few as ten to twenty words. This is roughly the number of snow words found in several modern languages in northern countries such as the United States, Canada, Norway or Sweden.

The origin of this controversy lies with one of the truly great researchers into the culture of Arctic peoples, anthropologist Franz Boas, who was born in Germany but did most of his work in North America. Boas studied the Inuit people of north-east Canada in the 1880s and, as anthropologists do, took part in their daily lives, learning their traditions and eating their food, which was mostly seal. He learned the language, and picked up on the nuances in the words for snow. The word *aqilokoq* meant 'softly falling snow'. *Piegnartoq* means 'snow where it is good to drive a sleigh'. Many other snow words followed the same pattern. In his *Handbook of*

American Indian Languages, published in 1911, Boas claimed that 'the Eskimos' had dozens, even hundreds, of words for snow. The anthropologists and linguists who followed made it a professional badge of honour to distance themselves from Boas. The claim was considered a speculative exaggeration, made to create a sensation and therefore, according to these truth-tellers, a myth. That people around the world continued to perpetuate this myth – including the very kinds of people I encountered as I was growing up, who knew nothing about Boas – only made it all the more important for critics to distance themselves from this unfortunate corruption of anthropology.

However, recent findings show that Boas apparently had it right, and maybe more so than even he realized. One of the more recent scholars to engage with the issue is another anthropologist, Igor Krupnik, from the Smithsonian Institution in Washington, DC. He has participated in several research projects in which terms for ice and snow in the Arctic played an important role. The amount of words for snow varies between different dialects of Inuit languages, but several have been shown to contain at least forty or fifty words. If you add the Inuit languages together, you get an even greater number of unique words in the whole language family. Krupnik himself is convinced that Boas only recorded words with their own distinct, individual meaning, and that his estimate of the total number of snow words was reasonable. The *Canadian Encyclopedia* also defends the position taken by Krupnik. In its 2015 online article 'Inuktitut Words for Snow and Ice', it presents both sides of the debate, including a list of the relatively small number of words for snow given in an Inuktitut–English dictionary. It then explains how some of the 'core words' apparently carry a number of other words with completely different meanings. The same is true for words about ice. They can also have distinctly different meanings even though they come from the same root.

I have already mentioned Yngve Ryd from Jokkmokk, who for several years collaborated with the elderly reindeer-herding couple Johan and Ibb-Anna Rassa, so he could compile the treasure trove of words in the Lule Sámi language that formed the basis of his book *Snö* (2001), in my opinion a work of immense value. The book contains 311 Sámi words for snow and related winter phenomena, such as seasonal vocabulary and types of ice. The latter are quite relevant to Ryd's book as a whole, which also deals with the necessary conditions of the frozen landscape for the traditional Sámi way of life and reindeer husbandry. But even if these words were removed, more than 250 remain that refer exclusively to snow. The nuance in the many concepts is dizzying, moving, even. Just why is that?

As every linguist knows, people coin terms for things that are important to them. These words bear witness to the breadth and understanding that takes place in people's lives that are lived in the presence of animals, nature, weather and climate. We can see it in virtually every word. The opposite is equally clear: most of us don't have a need for these words and probably wouldn't think we'd ever need them. Why would you and I need a word that means 'The reindeer gather together and go up to the snowfields to avoid the reindeer botflies'? That word is *tjussat*. Or the word meaning 'The snow gives way at the bottom. It's not the skis sinking into the snow, it's the whole snow that sinks' – *slamkedit*. It's striking that many of the words show the close relationship between the snow and the reindeer. Depth of snow isn't expressed in centimetres. That would be unnecessarily precise and you don't want to waste valuable time measuring. Instead, the Sámi talk about how far up the adult reindeer's hind legs the snow reaches. *Tjiebbemuohta* is snow that reaches the calf, after the first snowfall in autumn. The words reflect the concrete material reality.

More likely than there being too many words is that there are words that Ryd and Johan Rassa didn't include. Apropos the controversies in Canada over the number of snow words there, Ole Henrik Magga, a well-respected Sámi linguist, was asked how many words for snow there were in the North Sámi language in northern Norway, where he comes from. His answer was 180. I don't know if Magga has read Ryd's book. At any rate, the number of words is much bigger than the previous debates indicated. The important thing is to empathize and identify with the mental state of the Sámi.

Anyone who reads an account like the one I offer here will probably come to realize a few things. First, you should be careful about saying that something is either right or wrong. The most important thing is that all the nuances are present in the explanations and descriptions that are brought to light. This makes discussion of a subject easier and enhances the likelihood of eventual consensus on the matter. What's more, new research can change our perceptions at any time. What was once considered myth, can quickly become truth. Or the opposite – a given truth may turn out to have taken on a tinge of mythology. Just such a shift in polarity, or more accurately two – one in the 1900s and one in the last few decades – have occurred in the anthropological debate concerning the words for snow.

This can happen in any field. That's what makes knowledge exciting, like the snow crust in late spring as the sun rises. You glide marvellously along on a sled or skis, but you know that at any moment the sun may have warmed the thick snow and then you could be up to your belly in . . . well, a kind of white lie. *Tsiekkádahka* – 'before you sink with every step'. You should have stopped earlier. Or taken a different route.

Now for a third cliché, and one that has never left me: that snow is home to a far greater variety of living beings than just humanity. Thriving under the snow are all manner of voles and mice, crocus bulbs and dandelion seeds, beetles and spiders – all the things that come out during the short Nordic summer season, bursting forth to fill our meadows, pastures and forests, so that we can love summer, too. This cosy 'world beneath the snow' is created by vegetation that has been broken down and flattened by frost and snow, while the snow itself hardens into a thick roof. Out in the open it might be -30°C, but on the ground under the snow cover it can be above zero. In spring, you can briefly see where the mice and vole tunnels have run, before the green takes over.

Summer – the short, beloved Swedish summer – and snow. They're intertwined. Like everyone else, I learned about this in school. No one in the snow-covered part of the world could have failed to hear this evangelical message. It belongs to the same class of things as the Gulf Stream – which is something else that warms and delights our northern European climates – but on a smaller scale, something a bit homely. We learned that snow, even though it's cold, actually helps living things through the long winter. That snow is warmth, a nurturing blanket that wraps around small, fragile and valuable things like an angel's wing. Maybe snow has, after all, a truly critical role, and for that reason we can cope with it.

I would posit that most people in Sweden have heard variations of this belief. The truth of it isn't usually called into question; at least, I've never seen any evidence of it. I think most of us realize that snow is something we have to endure for the sake of the dormouse, the linnet and the viper. Otherwise we'd be living in a species-deprived taiga comprising only a few kinds of tree, such as spruce, aspen, birch and a couple more. According to the Swedish National Forest Inventory, which turned a hundred years old in 2023, Sweden has forty-five species of tree, twenty of which are native species. Most other countries have more than that, but we've learned to enjoy the ones we have, and if we were left with only half a dozen we'd probably be sorry. It would be unspeakably boring to live in a landscape with such uniform flora. So, we're happy to bear the yoke of snow, which is a light one, relatively speaking, and which also comes with a lot else to commend it, such as floodlit ski trails, snowmen, *lapphandskar* and moguls. Plus the silence, captured inimitably by Lars Gustafsson in the poem 'The Silence of the World Before Bach', which is in fact about winter and existence, and what you need in order to acclimatize to the total silence of winter, 'the skating silence' of a world before Bach touched it with his music.

But getting back to the matter at hand: where does this bright, snowy underworld come from? The subnival space. I'm asking the question right now, at this moment, during another winter, without knowing where it might lead. But before I discover a more prosaic answer, here's a thought: I believe that life under the snow is a kind of utopia, reminiscent of the way the romance of snow was used to rouse the Nordic nations. A romance that reaches right down under the snow. Because the life happening down there has its romantic aspects. It is a story of solidarity and care, of perfect communion and coordination. A perfect harmonizing of the seasons, which in fact have their roots in the universe and the solar system, and is accordingly a theme worthy of snow, which seldom shies away from the affective.

So, where does this notion come from?

First, there's something about cosiness. Swedish Wikipedia tells us: 'Thanks to the insulating properties of snow, the temperature is higher and the space is therefore a haven for rodents and other small animals.' We don't always use language like 'Thanks to' when we're talking about protecting rats and voles, but perhaps all is forgiven under the snow. 'Carbon dioxide levels can be high in the snow . . .' Is there a hint of regret in that tone? The alternative would be to see the snow as a national mousetrap – but perhaps that would be going too far. Fortunately for them, small rodents solve this problem by creating airholes through which to breathe. They also use them for foraging. A vole might come out to eat a couple of winter mosquitoes on the snow crust, gnaw a little bark off a bush or feast on seeds dropped by the trees. But there's always the risk of meeting a fox or a watchful hawk or owl. The fox hunts in slow zig-zags atop the snow, listening for the scratching sounds and squeaks of the desired small rodents underneath. It can detect sounds through more than thirty centimetres of snow. When the fox has the sound right underneath it, it jumps straight up and lands with legs outstretched and snout and front paws close together so that the layer of snow can be broken and its jaws can immediately fasten around its prey.

The Swedish Society for Nature Conservation is fond of the subnival. Back in 1969, more than half a century ago, it published the book *Animals, Plants and Winter*, and its campaign in 2023 confirms that the idea of nature thriving under the snow lives on. In 2018, *Sveriges natur* ('Sweden's Nature') published a description of subnival life so compelling that you'd have to be made of stone to resist becoming its eternal champion:

> Strands of old, chewed grass remain there, too, showing where the paths have run. [. . .] The lemmings must give birth to their young under the snow in order for them to be ready to quickly emerge in spring and create a peak lemming year. Since the nineteenth century, winters have become warmer, which is probably one of the reasons why big population surges have become rarer. [. . .] Snow fleas are not really fleas, but springtails. They appear as dark spots, a millimetre's length on the snow, that look like pepper or ash. They are so small that they can move right through the ice crystals in the snow cover. Inside, they can find food such as bacteria, old plant scraps, and ice algae growing on the snow. [. . .] The long, slender shape of weasels allows them to move along the paths under the snow, just as they do in the dirt paths in summer. Owls can hear the footsteps of voles under the snow, but they find it difficult to break through the thick snow cover.

Isn't that marvellous? We're learning about snow. And about the underworld of snow.

But this doesn't get me to the root of this concept. Where does the idea of a utopian existence under the snow come from? And when in history does it emerge? What does the *Swedish Academy Dictionary* – perhaps Sweden's foremost cultural infrastructure, founded more than 125 years ago – say? Now that it's online, I usually read it aloud during long family trips. After just a few minutes, the car or train compartment that had just now been as silent as the grave has transformed into a giggling, roaring seminar, feasting upon words we didn't know existed with such relish that no one is thinking about getting sweets any more.

But the *Swedish Academy Dictionary* provides no answers regarding the subnival. Nor on *nival*. Nothing to be found. On *nix* (snow in Latin) there's nothing either, except the given of course, the clear *nichts*. But under *nev-* (in one last feeble attempt), it appears: '*nevé*'! What's *that*? I don't know ... But as it turns out, it's a gold mine: 'Snow accumulation or snowfield in which the snow begins to turn into ice and which gives rise to a glacier.' Who'd have guessed? I fall to my knees in front of my desk out of pure emotion. I've learned something I never knew existed – the greatest of all rewards! 'Subnival' and 'nival' will surely be in the next revision of the dictionary, because these terms are on the rise even in Swedish, or that's my impression. Awareness is growing that these are sensitive environments at risk of disappearing along with snow. Think of the lemmings, the shrews, the springtails ...

Nevé sounds undeniably French, I immediately think to myself, and must come directly from Latin. Indeed, the word comes from the French *névé*, a word that evolved from *névi*, which has long been part of the dialect in the

French-speaking part of the Alps, areas such as northern Provence, Savoie and parts of Switzerland. This dialect version of the word took on a scientific appearance as early as the nineteenth century, when glaciological science began to emerge, but basically it is indeed Latin. *Neve*, without any accents, is the word for snow in both Italian and Portuguese, and it's now forever lodged in a cranny of my amygdala. My reptile brain recalls ski-adhesive tubes from Rode, Italy's top ski-wax brand, equivalent to Swix in the Nordic countries. '*Neve*' was written on each tube and on every jar of ski-wax. I always put Rode's 'Silver' in the wax box that I carried with me, a smart all-purpose climbing wax and a safe bet in difficult conditions.

The *Swedish Academy Dictionary* also makes reference to *firn*: German, more specifically from the Swiss-German *Firn*, meaning 'last year's'; related to the Old English *fyrn* and Old Norse *forn,* meaning 'before' or 'ancient'. So, it's old snow that has been lying in snowdrifts, gradually becoming compacted. Firn is very heavy, up to ten times denser than average winter snow. Firn snow is a circumscribed, albeit prolonged, phase in the various transformation processes that snow is always undergoing. These take place from the time snow forms in the clouds until it stops being snow, either by sublimating (evaporating, vaporizing) and escaping into the air, or melting (defrosting, thawing) and becoming run-off. Or it can become firn, the intermediate stage on the way to forming new glacial ice.

The world's glaciers are shrinking in volume and extent virtually everywhere, but firn is still being created all the time anyway, mainly at higher altitudes or in shady snow beds that are particularly sheltered, where snow can drift there in large quantities and linger over the summer. These are the places that the people seeking the Scottish snow-patches tend to visit, often in vain.

The first use of the term 'firn' documented in the *Oxford English Dictionary* dates from 1853; it's used by Elisha Kane, an American physician and scientific explorer. Kane was one of the oddly numerous proponents of the thesis that an open ocean existed around the North Pole. He even claimed to have seen this sea with his own eyes, which was more than most other people dared to do. Kane was also involved in the search for the ill-fated Franklin expedition, John Franklin's poorly equipped attempt to find the Northwest Passage to Asia, ending with the death in the snow of the entire expedition, all 129 men. An enormous tragedy, which saw both disease and cannibalism, to the horror of the Inuit.

In addition to some things I already knew, I discover in the *Swedish Academy Dictionary* that an early record of 'firn' in Swedish is to be found in the entry for 'Björling, Tyndall (1879)'. Who's that, then? It's not a difficult puzzle. Manne Björling was a professor of mathematics at Lund University who translated Irish physicist John Tyndall's key work *Heat Considered as a Mode of Motion* (1863) into Swedish. In the mid-nineteenth century, Tyndall was one of the people wanting to advance contemporary knowledge regarding snow and ice. He was very interested in the aesthetics of the individual snowflake and was well acquainted with the depth of research in the field. We've encountered Tyndall already as a reverent walker of glaciers and thinker about snow. However, he saw the snowflake as just one of the many pieces of the puzzle in understanding the frozen regions of the world. He sought something bigger and more important than just the single hexagon, however regular and beautiful it might be. After studying in Germany and getting a taste for real mountains, he soon became a keen and skilful mountaineer who annually went climbing in the Alps. Tyndall pioneered the transformation of mountaineering from being simply a sport, into the scientific study of high mountains, snow and ice, a development that took hold in exactly the time-period that the British Empire was the ideal arena, having glaciers on several of the continents where the Empire was present.

The word 'motion' for Tyndall meant the whole process of transformation of snow, from a falling flake to the formation of firn and new ice, as well as the rolling journey of the mature, giant ice as a sluggish river of what-once-was-snow to the glacial tongue, and its final melting-away. He wanted to take a holistic approach to snow and ice. 'Crystal' was a beautiful word, with wonderful overtones, but for Tyndall it was only a limited part of the truth. The great revelation in his work was that although snow and ice were on the ground, their form wasn't entirely fixed, and that the substance was never *still*, even though it looked motionless. The movement of the glacier, hard as pine, tough as resin, was a microcosm of the even greater movements involved in the mechanics of the truly great systems on and around planet Earth. Ultimately, everything is driven by heat, which was one of the many subjects Tyndall researched and subsequently wrote about – as he did atmospheric gases, such as carbon dioxide, and their role in the earth's thermal balance.

Tyndall's achievement was to begin to recognize the connections between these elements, gases and principles. And that it was all bound together by snow. The snow trapped the gases, slowed down the water, creating an intermediate state between heat and cold, a fine balance between earth, air and water. Not unlike what Antoni Dobrowolski would later call the cryosphere.

Tyndall performed successful experiments in which carbon dioxide functioned as a greenhouse gas, which meant that burning coal – and also oil, which had started to be used as fuel during his lifetime – would lead to an increase in the earth's temperature. What he glimpsed was the outline of what would become global physics. The term 'geophysics', a physics larger than the laboratory, encompassing both the planet and its surrounding atmosphere, can be found as early as the 1830s. But it was rare and didn't come into general circulation until the end of the century. At about the same time, as we have seen, glaciology also began taking shape as a discipline, with research on snow as a natural element. These things were connected. Snow wasn't just snowflakes.

In a way, Tyndall was working before the concept of geophysics had fully evolved. His vision was to develop a kind of physics on a planetary scale. But his concrete research and the great debates in which he participated centred on the individual glaciers, which he believed moved downwards, fed by snow that became first firn, then ice, a subset of the hydro- and atmospheric cycle of water. Others thought the glaciers flowed like sluggish rivers. Tyndall didn't think so. But they agreed that glaciers were in motion. American historian Bruce Hevly has called Tyndall's research 'heroic science into climate motion'. The adjective 'heroic' refers to the relationship that existed at the time between science and conquest: of colonies, of territories, but also of new terrain, what the motto of Britain's Royal Geographical Society called 'to the farthest ends of the Earth'.

Reaching things at any extreme was considered heroic, something that *real* men undertook. And while these extremes included deserts, jungles and savannahs, the most inaccessible and 'farthest ends' often involved snow.

It was a reflexive response to snow science's deep colonial root system that Gerald Seligman gave when, in a gesture that was more language politics than science, he rejected the eager Dobrowolski's proposal to call this whole scientific domain of cold 'cryology'. It sounded like 'crying', Seligman complained. He was very frank: it simply didn't make sense – as The Cure would later point out in their heavily ironic song 'Boys Don't Cry'. One line of snow research has undeniably continued on this theme, leading to a crescendo of angry debate in 2016 when a group of scientists in the United States, led by historian Mark Carey, wrote an article on the need for a 'feminist glaciology' (a subject we'll come back to).

John Tyndall himself wasn't a very typical representative for the geophysics of imperialism. We're already familiar with his aesthetic sensitivity to the snow and his openness to the mysterious beauty of nature. An Irishman with no trace of self-congratulatory boarding-school discipline and completely lacking in stiff upper lip, he was instead a restless and sensitive seeker, and a dazzling communicator of the new science to a public thirsty for knowledge, particularly London's poor and uneducated working class. But he was part of a system characterized by certain norms that were clearly achievement-oriented. 'Ever higher', as Dani Inkpen has called it in her study of the normative system of cold-weather mountaineering – the higher the mountain, the nobler the undertaking; and, as Vanessa Heggie has shown in her book on twentieth-century research in extreme and cold environments, *Higher and Colder*, the more potential prestige. A third book, *Suffering for Science* by Rebecca M. Herzig, visits a similar theme, that of sacrificing oneself for 'truth' by enduring snow and cold, among other things. Science isn't merely methodology and results. It's also a culture, a moral order, or more formally an 'epistemic community' – a community of people with similar knowledge, ambitions and practices. United, or sometimes blinded by, their project.

This is important. But we've travelled a long way from the utopia of the subnival microenvironment, with its small rodents and insects there under the snow cover. Instead, we've ended up in a totally opposite environment, on the windswept, exposed Alpine peaks with their nearby glaciers, among *firn* and *névé*, and men wearing leg wraps and fur trim. But perhaps that was the point of this chapter's spiralling discussion of words, etymologies, mountaineering and physics. The point is basically simple but, to be really meaningful, it requires us to follow Tyndall's lead and broaden our concept of what snow is.

For this reason, we should remember the word 'motion' in Bruce Hevly's comment. It refers to the movement of glaciers, as Tyndall contended. But snow is also part of larger movements, as an absolutely central element. It binds together the sky, land and sea in a mighty cycle. It unites the seasons. It exists at all altitudes, from the terrestrial layer, where it forms the roof of subnival space and shelter for the tiniest forms of life, all the way up to the top of the atmosphere, the roof of the whole of planet Earth, where it sets down its trillions of hexagons on the tops of the Andes, Pamirs and Himalayas.

Snow warms the ground and crowns the world.

Through similar vertical movement and exclusive zones, snow is also *social*. It's an element that can be found at the various levels of society. Snow unites and it divides. There's different snow for distinct groups of people. Some move in the higher spheres, using the snows of Gstaad and St Moritz, where they are flown in for expensive experiences and to be transported, vertically, by lifts, cable cars and helicopters, in a resource-intensive global fossil-fuelled tourism, the ultimate consequence of which is the very loss and solastalgia over the disappearance of snow that I described earlier in this book. Meanwhile, others continue to pursue their reindeer herding or hunting culture under increasingly uncertain conditions, often at somewhat higher regions above sea level, as is the case for the Sámi in the Nordic mountains. Still others, living or farming downstream, have their living conditions impacted, even devastated, as the snow drought spreads.

And still others, mostly those living their everyday lives at sea level – where the floods of the future will arrive – are trying to save ice and snow by various means. Geography is a recurring theme in Protect Our Winters (POW), snow's Western defence movement, mainly located in North America, Europe and the Anglo-American colonies. POW was founded in 2007 by American professional snowboarder Jeremy Jones. It wants skiers and boarders to take greater responsibility for their climate impact and for the snow sports community to become a stronger voice in the public debate. It argues that 'when individuals collectively choose to live sustainably, those individual choices can lead to a massively important ripple effect that can inspire policy changes and broad levels.' A core group exists among ski bums and elite cross-country skiers. Sweden also has a significant number of members.

The indigenous peoples of the cryosphere have their movements, too, the basic tenets of which are well captured by the title of Inuit activist Sheila Watt-Cloutier's 2015 book, *The Right to Be Cold*.

These hierarchies, the social and geographic differences, have all been affected by climate change. They're becoming evermore entrenched. Global divides are growing and, according to consistent analysis – ironically, much of it coming out of the World Economic Forum at Davos, the very town where the snow and elites are to be found – are making the world an increasingly insecure place. At the same time, snow's various interested parties are coming together. People who can speak with geophysical knowledge are now in ever closer contact with peoples that have been colonized and who still remain marginalized in many ways. In Sweden, scientists talk to reindeer herders, and also to skiers, hunters and fishermen who want to preserve the snow. They cooperate and change the narrative together. Swedish natural geographer Gunhild 'Ninis' Rosqvist works alongside Niila Inga, a member of one of the Sámi families who make a living from the Kebnekaise mountain region, and with several others in the Sámi village. Deeper understanding emerges from spending time and working together. These new epistemic communities also influence the public debate. Science-based judgement is gradually moving out of the camp of those exploiting natural resources and is now increasingly in collaboration with people who've drawn the short straw. People for whom change will mean loss. This shift has moral and political dimensions to it, just as epistemic alliances around resource exploitation usually have. But it's grounded in science.

In this sense, the subnival space is not simply an

earthbound utopia. It's also a picture of how differences can be reconciled in the presence of snow in order for everyone to live together. Snow is also biology, ecology, life. An inescapable part of a world in deep trouble.

I keep returning to the idea that snow is a transitional substance. A transitional element. There's life under the snow. There's life on the surface of the snow. There's life inside the snow. Life moves between these levels. In the atmosphere above the snow, birds and flies hover, feeding on the mice and worms that make their way to the surface. If the snow is allowed to rest for a long time, it eventually forms a solid body, ice. But even ice doesn't remain a solid body, with a single, definite shape. Like snow, it contains other bodies within it. It flows, moving on both sides of the border between the living and the dead. To call ice 'frozen' as though it means 'dead', as many have done, is a misunderstanding. It's probably more fruitful to think of the whole cryosphere as essentially viscous. Neither solid, liquid, nor gas, but something in between that unites all of those things and, in so doing, forms a natural part of the exchange between the earth and the atmosphere, with a different temporality than the water that runs quickly, and that falls as rain.

The slowing effect of snow isn't just a trait that can be used to store fresh water for agriculture or positional energy for hydroelectric power stations. Snow is part of a larger cycle, just as John Tyndall contended. Not only by transporting water between different forms of aggregation, but by being the most mobile element in the middle of the viscous sluggish flow through time and matter. And as a connecting substance between the solidity of the earth and the non-solid inconsistency of the air, from which a snowflake can be revived at any moment to be reborn in its infinite uniqueness when it comes into contact with something solid. Snow's temporality is also therefore of a different time from that of all the other elements. Viscous time is circular, with many possible speeds. An order of time for which no clocks yet exist.

So, when did the 'subnival' space come into our consciousness? It's about time I answered that question.

The term appears in the *Swedish National Encyclopaedia*, published in the 1990s, in a short entry of thirty-five words, among them the obligatory shrews and small rodents. In the 1955 edition of a previous encyclopaedia, the *Svensk uppslagsbok*, the expression is completely absent, and it doesn't appear in the 1917 edition of the *Nordisk familjebok* ('Nordic Family Encyclopaedia'), either. The subject of 'Snow' seems to have become generally less and less interesting to teach during this period: in 1917 the word is given a lengthy entry stretching over five columns; in 1955 it has shrunk to two; and by the 1990s to twenty narrow lines – little more than the space occupied by the entry for '*Snömögel*' (a type of plant disease). It's as if the twentieth century tried to unlearn everything to do with the material world and natural dependencies. The future was assumed to be insubstantial, weightless, artificial. I think this has changed now.

Whatever the case, the people who'd started talking about the subnival space, long before this expression came into use in Swedish, were naturalists or, often, nature lovers who were sympathetic to the Northern indigenous peoples. Canadian William O. Pruitt wrote in the late 1950s about the ecological properties of snow. He was also a passionate advocate of the scientific use of indigenous words, which he felt captured the ecological properties of snow much better than the conventional English words. The opposite of Seligman, in other words. *Qali,* a central concept for Pruitt, was the snow held in the tree branches that made the ground snow-free at the tree base.

Pruitt saw the properties of snow from the perspective of animals. His touchstone in life was a book that a Ukrainian colleague, William Prychodko, introduced him to as a young scientist in the mid-1950s. The author was the leading Russian winter ecologist Aleksandr Nikolaevich Formozov, and the book, which was his key work and published in 1946, was on the importance of snow cover for mammals and birds. Pruitt and Prychodko's ensuing translation into English was completed in 1966.

Formozov used the classical Greek word for snow, *khion* – from Chione, goddess of snow, daughter of Boreas, god of the north wind, and Oreithyia, lady of mountain gales – and divided the animals into three groups. *Khionophiles* love the snow and are evolutionarily adapted to it, like the musk oxen, snow leopards, lemmings, many voles and a wide variety of birds. *Khionophobes* avoid the snow as much as they can. *Khioneuphores,* including moose, deer, bison and other animals, can live in the snow, but not without issues. They get around with their long legs but are ill-adapted to deep snow and prefer to move about in areas with less snow, where they're less likely to be trapped by predators.

Formozov had discussed lemmings and small rodents that wintered under the snow, but a major breakthrough for subnival space came in the early 1960s. German researchers studied *der bodennahen Luftschicht* – the layer of air close to the ground. The Swedish biologist Carl-Cedric Coulianos wrote about 'microclimates'. Olle Eriksson from the Swedish University of Agricultural Sciences held courses in the 1970s about the snow crust and what was hidden under it. At the Rickleån River in Västerbotten, entomologists from the newly founded Umeå University caught winter spiders and insects. They buried pit traps containing fatal glycol in autumn, covering them with a Masonite board to prevent the creatures' escape and keep out the snow. When the snow fell, they accessed the insects trapped in the subnival space via a shaft they had previously prepared. Small-mammal researchers did the same in scientific experiments with voles and mice.

Once the empirical groundwork was laid, the basic features of the perfect story were ready. For the Swedish Outdoor Association, the Scouts and the rapidly growing Swedish Society for Nature Conservation, the image of a secret world to be discovered under the snow was an ideal one. Incidentally, Coulianos was the founder of the Field Biologists, the advanced youth chapter of the Nature Conservancy, and also its first president. Maybe you shouldn't look too deeply into this magical world – you might risk damaging it – but just knowing it existed was enough for most! Children's imaginations did the rest. Schools did their part, too. Nature under the snow was part of seasonal awareness, important in a seasonal country, and became the first new addition to the topic of *hembygdskunskap*, which explored the relationship between nature and the community, since it was added to the school syllabus in 1919. It was still on the curriculum when I started primary school, in 1963.

In this ecological idyll, the sweet little shrew fits right in. As does the cute, plump vole living side by side with the worm and the crocus, as the lion and the lamb once did in Paradise. Because that's what this is: a child-sized Eden made of snow. And shame on anyone who thinks negatively about it.

It's just that we now know all the other stuff, too, about the vertical politics and movement of snow. And how these big movements are connected through the snow itself.

Nivea

*

WHERE I GREW UP, IN AN AREA RICH IN SNOW, not far from the Arctic Circle, few people were suntanned in the winter. The sun shone low, and it took reflection from the snow for the sun's rays to be able to take hold. Only faces and hands were exposed. Otherwise, we assumed that all of us had skin that was more or less pale. Numbering among the more tanned were the people doing forestry work. They stood all day in the snow, and even in the shadows from the trees there was still the light reflected by the snow. Then there were the Sámi, whose winter feeding of reindeer meant they were frequently outside. Other than that, there were some outdoor enthusiasts, who were often teachers, and then the odd individual who'd travelled 'down to the heat' – that's how we referred to Mallorca and other places down in the Mediterranean. It might be some person who worked in a bank or pharmacy who was tanned in the winter, possibly all the way up their arms, and it didn't really make sense until you considered that maybe he or she had been 'down to the heat'.

Coming into spring, they might have gone over the Easter holidays. Meanwhile, a lot of people went to the mountains then, fishing for char. Not our family. Not until my grandfather built the house in Klimpfjäll, at the corner between Lapland Jämtland and Norway. I was part of the construction. Then we, too, were tanned in April. I was seventeen and in high school; my youngest brother was three. So I had been pale in the spring fourteen years longer than he had at that time. I would go skiing in the mountains. That's when you get tanned, whether the sun is shining or not. It's one nice thing about snow – you can get a tan quickly. But . . . it can also be a little dangerous.

The worst thing was the eyes. It started when I was doing a lot of longer-distance skiing in my teens. Then my eyes hurt, especially when it was snowing, and the worst was drifting snow. But that then passed. When I became a senior at twenty, I skied longer races. But a 50-kilometre race in Lycksele in January, in -20°C and heavy snowfall, wasn't healthy. I came third, so athletically I was satisfied, but afterwards I had to be rushed to the hospital – I couldn't see at all, and my eyes were burning. I had scratched my corneas, the doctor said. I was given a topical painkiller cream and had to sit quietly at home doing nothing.

They likened it to snow-blindness. But it wasn't the light, it was the abrasive crystals in the snow. Sharp needles from the grey sky, there because of the cold. No *lapphandskar* could form in that kind of cold. It seems that I had sensitive eyes. I started experimenting with swimming goggles and slalom goggles, but they would fog up. There were no proper skiing goggles at that time. I took a chance with it again, and the same thing happened. At least now I knew what it was. The doctor told me to be careful because there might be scarring of the cornea, and then my eyes might suffer refractive damage.

I gave up competing, but continued to go on extended

treks in the mountains both in summer and winter. After a ski trip in May during intense sunshine, I experienced real snow-blindness. The same symptoms, only even more excruciating pain. It was early summer in Umeå. I lay in a darkened dorm-room having to supplement the pain medication with whisky – that's how painful it was. I was afraid I could have lasting visual issues. After that, I got glacier glasses with leather rims that completely enclosed the eye socket. I always carried them in my rucksack. Even in summer, I needed them if I had to walk on a glacier for any length of time.

Skin could also be damaged if you got too much sun and snow all at once. You'd get burnt. For that, there was sun oil and sunblock. I had a sensitive nose – narrow, straight and long, with the worst possible angle to it. I burned it several times until I realized that it needed to be protected. I put on bandages or tape because ordinary sun block wasn't enough. Eventually I discovered zinc paste in a Mediterranean bathing shop. Mine was pink. I looked insane, but it gave complete coverage and I didn't get sunburnt any more.

I could have avoided some of these problems. It was certainly well known that sun and snow damaged both eyes and skin. At the museum in Umeå, I'd seen the exhibit on seal hunting in the Gulf of Bothnia. A few generations ago the hunters would approach the animals by lying horizontally and stealthily sliding on their wooden kick-sleds behind camouflaged screens of white canvas that hid their faces and blocked out the worst of the light.

Sun protection for skin didn't come into being until some sunbathing pioneers wanted to expose their whole bodies, which became fashionable in the 1920s. Swimsuit coverage shrank rapidly after the First World War; taking off anything that was bulky and concealing was in vogue. So was being tanned. One of these pioneers was Coco Chanel, who was soon to create the women's fashion of the time with straighter cuts and distinct angles. This communicated forward thinking, an appetite for life. Previously, the upper classes had preferred the shade, hiding themselves indoors, both men and women enveloped in modest, fully-covering clothing. The only reason anyone had a tan was because they'd been doing physical work outdoors – this was still the case in Åsele when I was growing up. Coco Chanel and the 'flappers' of her time turned the conventions of the period upside down.

The relationship between snow and modernity unfolded with the body as the stage. Snow can be felt; it is cooling, moisturizing, reflective. Snow isn't just a material, it's a medium, too. There was nothing that could effectively stop its impact. People wanted to be outdoors. 'Outdoors' becomes an important word. In some countries, the word takes on an overtly positive connotation. Again, Fridtjof Nansen comes to mind. Norway's hero is often depicted outdoors, on skis. One of the most iconic images of him captures the moment when he and his fellow explorer Hjalmar Johansen have left *Fram*, their ship that had frozen in the pack ice, to try instead to reach the North Pole on skis. They don't make it, and on their way back south they're forced to spend the winter on Franz Josef Land, where they miraculously meet English explorer Frederick Jackson and are rescued. Jackson pulls out his camera. Nansen looks feral after fifteen months in the snow, his eyes wild and white, alien-like. The photo isn't a close-up, but his skin is much the same colour as his dirty clothes. It's easy to imagine how imbued it is with bodily substances – hair grease, dandruff, sweat, salts – plus cod liver oil and seal and walrus blubber. No sun block, but still a high sun-protection factor. Underneath it all, pigment activated to its full potency. A skin in keeping with his highly erect, virile pose in the photo, positioned in counterpoint to the skis that are under his feet.

A copper-skinned Nansen, touched by divine grace. Outdoors. That was 1896. Perhaps his skin itself, his feat and the miracle of his rescue, can all be seen as signs of a pattern: steps on the road to independence from Sweden. An outdoors nation must be led by these kinds of survivors. The following year, his Swedish counterpart, the pale engineer Salomon August Andrée, disappeared off the horizon after his hot-air balloon crashed in the same polar sea ice where Nansen had abandoned *Fram*. Andrée's remains were found in 1930 on White Island, not far from Franz Josef Land. It's still unclear how he died, but no one thought of Andrée's skin as particularly interesting, or of the snow's abstract modernity to the aircraft voyager. Andrée, ever the optimist, was adamant that his technology would work and that he would float smoothly above it all. He didn't, and by the time he made contact with the ice and snow, he was lost. Nor, it seems, was he quite so optimistic after all; it has not gone unnoticed that André carried a pedometer with him.

For thousands of years, people have been taking care of their skin, but before the First World War there wasn't a product with sun protection for skin that came into the contact zones of air and snow. These elements met skin in almost exactly the same way that the wooden underside of the skis met the snow: another zone of close and intimate contact. On skis, this space was covered with wax, a product that would eventually become a subdivision of the chemical industry.

Here were two modern gaps that needed to be filled. But they weren't filled following a straight line. The story of the modernization of snow leads in several directions at once. There are so many new beginnings at this point that it's almost impossible to keep track of them. Yet it is possible to discern certain patterns. In any event, the concept of snow we have at the end of the Second World War is quite different from the snow we had at the end of the First. Snow is no longer the exclusive realm of manual workers outdoors – snowshovellers, reindeer herders, lumberjacks and farmers who transport their goods by horse and sleigh along the Norrland coast, with steamboats and railways yet to be introduced to the region. Instead, it is becoming something of interest to the elite, and for people in general. If the word 'revolution' can be used to describe snow, it's here, in the interwar period, that such a revolution takes place. Except the revolutionaries don't know each other, and they don't know that what they are involved in is a revolution.

A good place to start tracking this revolution is at the top of a mountain delineating the border between the canton of Graubünden in Switzerland and the state of Vorarlberg in Austria. Vorarlberg borders not only Switzerland, but Germany and Liechtenstein, too. The mountain, which is 3,312 metres high, was first scaled in 1865, when large parts of the mountainside were covered by a continuous glacier. Today the glacier has melted and fragmented into smaller sections.

The mountain was also once the border between free

Europe (Switzerland) and Europe controlled by the Nazis. In 1936, a group of devout Christians had a cross carried to a plateau at the top of the mountain, much like Christ on Calvary. The *Vorarlberger Volksblatt*, a Christian socialist newspaper, described this act as a 'sign that this country is and will remain Christian despite all the attacks of the "destroyers of Christendom"'. We call this a symbolic act. Symbolic acts are, as a rule, the most important things we undertake. I find myself trembling when I think of this cross-carrying taking place in the same year that Hitler's Germany was organizing both the Summer and Winter Olympics.

The name of the mountain? Piz Buin, which is Rhaeto-Romanic for 'ox head' (and pronounced 'pitts bwahn', with emphasis on the second syllable). The mountain looks like an ox, with a broad bull's neck at the top. Two years after the cross was brought up the mountain, Swiss chemist Franz Greiter summitted it. He was badly sunburnt during his ascent, due to the sun reflected by the snow on the mountain, and in the following years, in collaboration with his wife Marga, Greiter developed a skin cream with a sun-protection factor. It went on sale in 1946 as 'glacier cream' (*Gletscher Crème*) and was named Piz Buin after the mountain where Greiter got sunburnt. Today, it is still one of the world's strongest brands of sunscreen and sun oils.

There had been previous attempts to protect the skin from the sun's rays. In the 1930s, experiments with zinc oxide particles blocked sunlight effectively. At the same time, it was important not to block out all the rays, so the skin would still bronze but not burn.

Nivea, the dazzling-white skin cream, was a sensitive case. The name itself comes from the word for snow, and means 'snow-white' to be specific – as in the bird *Pagodroma nivea*, or 'snow petrel' in English, of the storm petrel genus. It is one of only three bird species recorded at the geographic South Pole. *Pagos* is Greek for 'frost' or 'sea ice', *dromos* means 'runner', thus 'snow-white sea-ice runner', because it looks like it's running on the water when it takes off. 'Snow white' probably makes more people think of the old fairy tale with roots in the Middle Ages that the Grimm Brothers published in 1812: *Schneewittchen und die sieben Zwerge* ('Snow White and the Seven Dwarves'). In Latin it would be *Nivea et septem homunculi*.

I had never thought of *nivea* as a word associated with snow, but only as the Nivea brand. Even though I'd grown up with it, in the middle of all the snow. Mother used Nivea and made sure we kids had the cream on our noses and cheeks whenever we went out in the sun. Especially in late winter, for skiing trips, or when we followed Dad to bait the nets we'd put out for pike under the ice on our lake in April, when pike are best. Mother was most careful with herself. The thick, white cream contrasted gently and sportingly with her big brown eyes as she massaged it into her soft lips. She looked like a model, leaning gracefully against a pine tree in the spring sunshine, and might have served as a modern role model in inner Västerbotten who in some quiet way evoked the movies and glamour. For a long time, I didn't know of any other skin cream, and the round blue jar with the white text was a family symbol that I associated with outdoor life and freshness. Also with thoughtfulness.

What I've written here is all I knew about Nivea, and the memory is fleeting. I hadn't even reached puberty before I left Nivea behind, forgetting all about it – until 2011, when I stepped into Nivea's huge flagship store in Berlin and realized the true depth of this thin, flat jar. The now world-famous brand was celebrating its centenary. In fact, it was only then that I began to realize how skin and the care of it were also incorporated into Third Reich propaganda. And that one of the connection points for them was the snow.

Nivea was, and still is, part of the Beiersdorf Group, which has been manufacturing plasters (Leukoplast and Hansaplast) and skincare products since 1882. In 1890, the founder, Paul Beiersdorf, sold his creation to Oscar Troplowitz, a Jewish pharmacist and businessman, who quickly and successfully grew the company with new products, including Nivea. Several members of the management team during the interwar period were also Jewish. At its core, Nivea was a modern product with a modern aesthetic. The target group was initially the white, delicate *femme fragile* of the Art Nouveau. During the more open and international 1920s, the brand changed its style.

The scientifically developed cream was designed to help women become both attractive and independent in their professional and public roles. The Nivea woman was sporty, she played tennis, cycled and skied, and was often portrayed in ads in a wintry setting. All-German models with well-nourished appearances were recruited in the form of three young ladies from the Lübeck area. Business boomed.

When the Nazis came to power in 1933, the company came in for criticism from its German competitors and the anti-Semitic press, which accused the group of not being Aryan enough. Fearing that business would suffer or the company would be seized, management hastily decided to move its already extensive international operations to Amsterdam, where it could act with more independence, maintaining the confidence of other countries in which it did business. All key Jewish staff left Germany. At the same time, 'German' Nivea remained at home, adapting its aesthetic, language and advertising to the requirements of Hitler's regime and allowing the authorities to control the details of its corporate culture.

The company's image now became an integral part of the 'traditional family values' domestic propaganda prescribed by socially conservative Nazi ideology, which Beiersdorf zealously tried to fulfil, precisely to avoid being associated with anything Jewish. The sportier look of the twenties worked reasonably well but was altered in an Aryan direction. Models were drawn as blonde, sometimes wearing lederhosen and dirndls. The ideal woman was no longer shown as slim and cool, but more like a working mother – sporty, healthy, natural, and free of seductive make-up, all at the same time. The typeface used in print advertising was now the regime-imposed Fraktur style, a clear departure from the modernist typeface that had previously dominated the brand.

Elly Heuss-Knapp was an important creator of Nivea's radio and cinema advertising (until this was banned by Goebbels' Propaganda Ministry in 1935). She freelanced for the company and used both snow and skiing in her campaigns. When the Winter Olympics were held in Garmisch-Partenkirchen in 1936, the Nivea advert *Weiss in Blau* ('White in blue') was shown in cinemas. You can see it today on YouTube. 'Proper' German families in stylized animation are shown on the beach, on a hike in the mountains, and ultimately on skis and toboggans in the snow, making flowing and fluid progress to the tune of upbeat music.

Another animated Heuss-Knapp advert, *Katharine*, shows a female office worker keeping in shape at the gym. She is praised by a Nivea advertisement designer, who heralds her as the ideal for his Nivea poster, the perfect embodiment of a young German generation meeting the regime's demands for health and racial purity. Outside Germany, the adverts could be focused slightly differently, but snow was preferred in those, too. In 1941, a scantily clad Nivea model poses for the French-language market in front of a mighty Alp, her ski pole visible and her hands in knitted mittens in stark contrast to her bare shoulders.

Politically, Heuss-Knapp's views were a far cry from the Nazi regime's ideals. She had stood as a liberal politician in general elections and been banned by the Nazis from making political statements after the 1933 *Machtergreifung*. Her husband, Theodor Heuss, also a liberal, became the much-loved first President of the Federal Republic of Germany, in 1949. There's nothing to indicate that Heuss-Knapp was a traitor; nor was Beiersdorf unique. Several German brands (Hugo Boss, Mercedes, BMW, Bosch) were associated with German elite circles before, during and after the war. Adherence to the regime, in Beiersdorf's case even to its aesthetic ideals, didn't necessarily mean a similar ideological conviction, at least not for everyone. But it was necessary to be able to recognize what the prevailing norms were, and to have made a moral calculation. Beiersdorf had chosen opportunism and the prospect of survival, instead of confrontation and likely seizure of the company.

Snow and winter sports were important to the Nivea brand. Skin was impacted by the sun, and the impact was multiplied by wind and the reflection of light caused by snow on the ground. Nivea protected 'against the biting cold and wind' – '*contre les morsures du froid et du vent*', as the 1941 French advertisement put it. It was during the inter-war period that Alpine sports started to take shape. Lifts and cable cars were built to take skiing into new places, expanding the concept of healthy outdoor activity that had been tried earlier in the original winter sports resorts. The growing number of people who sought snow-covered, sun-drenched terrain in this way became an important target market for Nivea, and so good for business that Nazism, like Italian Fascism, found snow to be a perfect element for ideological propaganda. The war itself also made its contribution: the blue Nivea tin was part of the Wehrmacht's equipment and strengthened Beiersdorf's position in the market.

Snow was to be found in the mountains, and mountain landscapes were an integral part of the aesthetic cult of the Nazis. Mountain films were already a popular, even cherished, genre. Leni Riefenstahl made her acting debut in 1926 with the film *Der heilige Berg* (*The Holy Mountain*), bringing this experience to her later role as a propaganda filmmaker in the Third Reich. Riefenstahl also acted in other films by the same director, Arnold Fanck, such as *Die weisse Hölle von Piz Palü* (*The White Hell of Pitz Palu*) from 1929. While working in 1932 on Fanck's *SOS Eisberg* (*SOS Iceberg*), she spent four months living in a tent in Greenland. In another Fanck film, *Der weisse Rausch* (*The White Ecstasy*, 1931), Riefenstahl played opposite the Austrian ski virtuoso Hannes Schneider. He had previously appeared in Fanck's 1920 film *Das Wunder des Schneeschuhs* ('The Wonder of Skis'), which is said to have been shown in Stockholm as early as 1922 and was shot at an altitude of 1,800 metres in Schneider's own 'ski pedagogium' in

St Christoph, Tyrol, where he trained skiing instructors. In the film and in the book of the same name, Schneider taught a new technique called *geduckte Stemmkristianiatechnik*, also known as *tiefe Arlbergtechnik* (Alberg technique), which involved using greater vertical leg movements. According to a report in *On Skis*, the 1926 yearbook of the Swedish Ski Association, the days of the Norwegian telemark turn were numbered.

The fates of Schneider and Riefenstahl illustrate how different individual careers in the snow could become, depending on one's politics and morality. Schneider was an anti-Nazi and was imprisoned after Germany's annexation of Austria in 1938, but in 1939, with the help of the Americans, he was smuggled out to New Hampshire and Cranmore Mountain Resort, where he trained a US Army division in skiing during the war. He remained true to his Arlberg technique and ideals of freedom, embracing the snow as a playful surface for downhill skiing. Leni Riefenstahl, meanwhile, became Hitler's propaganda-film maker, with a special talent for inflating and sanctifying the power of the dictatorship and a presumed community of the people under the guise of 'beauty', in accordance with what the philosopher Walter Benjamin referred to as fascism's 'aestheticization of political life'.

In this role, she would return to snow, continuing to use it as an aesthetic and political instrument, but now as a medium for her increasingly ideological orientation. *Das Blaue Licht* (*The Blue Light*, 1932), which she directed, is set in the Dolomites. Then, in February 1936, she's featured on the cover of *Time Magazine*, photographed from below wearing a bathing suit in a wintry Alpine landscape, skiing upward towards higher altitudes. The angle of the picture had implications, the headline reading simply, 'Hitler's Leni Riefenstahl'. The accompanying report about the Winter Olympics, which were taking place at the time, mocks Riefenstahl's careerism, her friendship with Hitler and his soft spot for the former dancer. It was known already that Hitler had commissioned her to make a film about the upcoming Summer Olympics in Berlin, unsurprising given her success with *The Triumph of the Will*, which showed the Nazi party's events in Nuremberg in 1934. *Olympia* was released in 1938 and, despite political ambivalence in many quarters, brought Riefenstahl to Hollywood. There was no such ambivalence in other quarters, however. Torsten Tegnér, the politically evasive editor of the leading sports newspaper *Idrottsbladet*, was quoted enthusiastically in the Nazi press announcing that Leni Riefenstahl had received Sweden's Porla Prize (founded by the Porla Brunn mineral-water company) for 'activities beneficial to Swedish sport'. Riefenstahl was convicted of complicity after the war but maintained her innocence, claiming that her aim had simply been to create 'beauty'.

Snow could lead in different directions. Snow in the interwar period was just that: ambivalent, controversial. It became increasingly important, as it was enjoyed by more and more people. It was at this time that ordinary skis became a regular commodity filling sporting goods shops, which in Sweden, for instance, became common even in the smaller towns and villages. Sometimes the shops sprung from ski manufacturing enterprises, like Alewalds in Stockholm, which grew out of Sandström's ski factory in 1937, or they were started by sports enthusiasts – for example Martin's

Vapen & Sport ('Weapon & Sport') in Umeå, owned and operated by cross-country skier 'Gold Martin' Lundström, who
took double gold at the 1948 Olympics in St Moritz. Skiing
factories popped up in many countries, often demonstrating great local variety. The spring holiday known as 'Sports
Week' started at this time in Sweden, with school trips to the
mountains. The concept of snow ventured to another level,
and into other places, which then became designated snow
resorts in Sweden: Åre, Riksgränsen, Storlien and Hindås,
to name a few. New infrastructure began to take shape
for lives into which some indecision was creeping; maybe
there was no place like home, but ideally you might also get
away from it from time to time. The snow became one such
attractive place to get away to. Propaganda was a perpetual
presence. Snow fostered health. Snow provided education
and discipline. Dealing with snow required new knowledge
and a new skill set. Instruction books began to be published.
Storm warnings were issued. Avalanche dogs were trained.
And the army becomes a major purchaser of skis.

It was also then that competitive skiing as we know it
began to take shape at an international level. The International Winter Sports Week had been inaugurated in
Chamonix in 1924, and was subsequently raised to the status
of being the first Winter Olympics. Norway won most of the
events that took place on snow. The next Winter Olympics
were held in St Moritz, in 1928, followed by Lake Placid, USA,
in 1932 and Garmisch-Partenkirchen, Germany, in 1936. The
Olympics themselves were politicized; they created dilem-
mas like the ones companies had. Was Nivea in favour of the
emancipation of the modern woman or for the victory of the
Third Reich? It was hardly possible to be both at the same

time. What was to be done when international competitions were organized by politically problematic regimes?

The 1941 World Ski Championships held in Cortina were organized jointly by the Axis powers of Germany and Italy. They were later invalidated and no longer appear in the medal record books. Few countries took part, but Sweden was one of the ones that did. The 1942 edition of *On Skis*, the yearbook of the Swedish Ski Association, included an enthusiastic travel report from the successful championships, written by the downhill skier and ski guru Sigge Bergman. Another article, by Gösta Wetterhall, related the efforts of Swedish military athletes, which he deemed brilliant. In retrospect, the decision to take part is difficult to justify. Sweden's participation was bad enough, though it is probably explained by the fact that in 1941 the Axis powers Italy and Germany could still be perceived as successful and the Swedes thought they were in honourable company. The Norwegians, who of course did not participate and couldn't fathom the Swedes' priorities, wanted Sweden to ask the International Ski Federation (FIS) to annul the championships after the war. Surely that was reasonable? But the Swedes held fast to their medals and refused. They were even less inclined to apologize, and the feud between the two nations' ski federations continued throughout the autumn of 1945 and well into the winter of 1946. Eventually the Swedish athletes gave in, but not before the Swedish press, who initially thought the Norwegians were petty and stubborn, had changed their minds about that.

The Swedes' lack of judgement had consequences. Swedish army colonel Carl-Gustaf Hamilton, who was also the author of the 1926 book *Från det vintriga Lappland* ('From

Wintry Lapland') and a member of the patriotic Lapland Mountain Men's Club, had to resign as chairman of the FIS. Hamilton had been vice-president, taking over on his own initiative when the incumbent president, the Norwegian Nikolai Østgaard, was prevented from serving for understandable reasons: he had spent the war years as King Håkon's aide-de-camp with the Norwegian government-in-exile in London. Not content with replacing Østgaard, Hamilton also applauded the Axis powers' spectacle held in Cortina. To show their appreciation, the Germans made him a 'Commander of the Order of the German Eagle', instituted in 1937 by Hitler himself and awarded until the fall of the Third Reich in 1945. The United States, on the other hand, withdrew from the FIS, accusing Hamilton of direct collusion with Germany. Swedish skier Martin Matsbo, who'd competed in Cortina, was banned for several years from the FIS's annual Holmenkollen games, in which he'd won the 50-kilometre race in 1937.

Another career, that of Officer Enrico Silvestri, serves to illustrate how difficult it can sometimes be to decipher what's been politicized. In December of 1995, I bought a book in an antiquarian bookshop in Milan written by Silvestri: *Lo sci agonistico: Allenamento e gare* ('Competitive Skiing: Training and Competitions') published in 1943. I was at the time a representative in the European Commission Visitors Programme, one of two Swedes, usually politicians, diplomats or intellectuals, who were selected annually through opaque processes. It was the final year Sweden had participants, because it became an EU Member State that year. I had chosen to focus on regional development in northern Italy for my project. I had also, so far mostly at random, started collecting skiing literature for a book I was planning to write (it eventually became *Kroppens geni*, 'Wisdom of the Body', published in 2011). But I didn't use Silvestri's book, other than for a short comment about what I perceived as 'educational ski fascism'. Is that what it really was, though? Rather, today I see the book as representative of the ambivalence and growing tensions around snow that became apparent during this period.

Silvestri was from Turin, in the foothills of the Piedmont Alps, and had studied law in his home town. He was a talented skier, competing with military teams in two Olympic Games as he rose through the ranks. At the 1936 Garmisch-Partenkirchen Olympics, during Mussolini's regime, he led the Italian squad that won gold in 'Military Patrol'. The International Olympic Committee didn't want the event to be in the Olympic programme, but the Germans pushed for it – Hitler got personally involved – and managed to squeeze it in as a 'demonstration sport'. The winning Italian team was celebrated throughout the country and each participant received 30,000 lire from *Il Duce* himself. Both the Garmisch-Partenkirchen Games and this competition, advocated by Hitler and accordingly rewarded by Mussolini, can be seen as part of a rising militarization of snow.

Silvestri's own personal development, however, was to take a different direction. During the war, he wrote his 350-page instruction book, *Lo sci agonistico*, which was quite wide-ranging, with lessons in all aspects of ski sports from ski jumping and slalom to glacier climbing. The book has a wealth of black and white photographs. They are remarkably instructive, partly because a middle-aged male model is shown in elegant, white Roman-style underwear that allows details of body movements to be clearly discerned (though it can't be Silvestri himself, since he is also listed as the photographer). Silvestri was also involved in the resistance movement against the German occupation, after Mussolini was deposed and murdered, and fought against the Germans on the Allied side from January of 1944 until the end of the war. Books, or people, are not always what they seem.

German movies featuring glacier cathedrals were joined by films about expeditions, which received large amounts of funding and were intended, among other things, to show how there was cooperation in the research being carried out to deepen insights into the ancestral Aryan homeland. In Nazi ideology, the highest mountains were surrounded by an aura, as were the animals associated with them, such as birds of prey, especially eagles. Italian fascism worshipped similar ideals and launched its own self-aggrandizing projects with expeditions to the Karakorum and Pamir. Similarly, the image of the alpinist, the mountaineer, became revered; a lone climber on the way to the purity of the highest peaks, an athletic ascetic with the strength the Nazis admired.

The mountaineer became a key figure in the exploration of the primordial Aryan home, the *Heimat,* and well-equipped expeditions were sent to the Himalayas, where the highest peaks on earth were to be found. The obsession with the highest peaks was rooted in the fascist consciousness, in which nature comprises a hierarchy. At the very top stood the *Übermensch,* who by virtue of his biological purity and power had a mandate to lead the world, and the greater right to live in it. It was according to that same logic that aristocratic animals such as the eagle, the moose, the bear and the bison (which Göring admittedly trophy-hunted in Poland's Białowieża Forest) deserved to be protected, and landscapes and waterways were considered the life-giving soil and blood of the nation.

Snow as an element could be inserted into this hierarchy of values. On the highest mountains lay the eternal snow, a wreath of cold that testified to the immutability and permanence of the characteristics peculiar to the Nordic countries. In this sense, the earth itself had a hierarchy, that was essentially climatic. The best traits were developed by people living in cold climates, where the Darwinian struggle for existence was allowed to reign unchecked. Cold toughened a person. That's why the Nazis also had a cult of the Nordic region and Nordicism, where the cold was strongest. At one time this belief was commonplace. 'To Aryan blood, the purest and oldest, / to Swede I was married by a friendly norn', wrote Swedish poet Viktor Rydberg in 1886, heralding what he perceived to be an act of grace by a benevolent destiny.

But neither linguistics nor archaeology could back up the fantasies of 'Aryan ruins' among the snow deserts of Asia. What happened instead was that while such ideas were abandoned as unrealistic, they found a second home with the xenophobic racists who needed them to legitimize their increasingly crude expulsion campaigns of 'subhumans' such as Jews, Roma, 'deformed' people and communists. Nazi ideologist Alfred Rosenberg declared that the most unspoiled, racially pure Aryans, who had been subjected to a minimum of racial mixing and were the noblest heirs of the Indo-European ancestors, were in Scandinavia. Similarly, lowland people from warm climates were of lesser stock. Snow and ice became high and noble elements, and places where the harsh laws of this natural whiteness prevailed were surrounded by the same aristocratic nimbus.

However, this couldn't be translated into a simple political geography. Germany itself was neither Arctic, nor particularly Alpine. Nor was it particularly Aryan. All peoples were mixed, even Rosenberg had to admit that. Admittedly, Nazi support was strong in the southern German Alpine valleys, and the ski-resort town of Oberstdorf has been singled out as the place in Germany where the Nazis had their strongest popular support. But states were not racially perfect creations and this was not a precise demographic doctrine. Rosenberg's widely disseminated magnum opus, *Der Mythus des zwanzigsten Jahrhunderts* ('The Myth of the Twentieth Century', 1930), is, like most of what he published, one big conspiracy. 'Myth' is the operative term, and he himself invented the concept of 'subhumans'. In the Nazi imagination, it was through the mythical connection with the original Aryan home – in a vague place in the mountains of Central Asia – that the longing arose in the soul of the people to restore what had once been lost to harmful racial mixing, social levelling and cultural decay.

Snow and ice were a curious link. SS chief Heinrich Himmler, deeply influenced by pagan cults and mythical ideas partly inherited from Alfred Rosenberg, was a staunch supporter – along with Hitler and others – of the so-called *Welteislehre* ('World Ice Theory'), invented by the Austrian engineer and inventor Hanns Hörbiger. The main idea of this doctrine was that ice was the foundational element of the world, something that permeated the development of the universe, and that it continued to be so. One interpretation of it, embraced by Himmler, was that the Aryan race had emigrated from Europe and then made their home in the high elevations of snow and ice found in the Pamirs or

Himalayas, and that Aryan descendants formed the social elites of much of Asia. Actual expeditions were sent out at the end of the 1930s, with experts from the cultural heritage foundation, the Ahnenerbe, created by the SS chief to look for archaeological remains of this alleged high culture that would prove the Third Reich's historical designation as the chosen people of World Ice Theory.

Of course, there was no evidence to be found, but this type of performance contributed both to the cult of the North in Nazi thought and to the high status of snow, ice, cold and the highest mountains in Nazi geography. With this cultural–political baggage, it wasn't surprising that Germany made some attempts to assert itself in polar research in the 1930s and 1940s. An expedition to Antarctica in 1938–39 was driven primarily by a German interest in whaling in search of fat for margarine, which needed to be laid away for the war effort. But some research was also carried out at the same time, indicating intentions to take up the fight for the White Continent. Aerial photographic material taken by the Germans was published in 1942 and attracted some interest, for example from the Swedish glaciologist Hans Ahlmann – an ardent anti-Nazi – who thought he could use the images to discern traces of the same warming in the Antarctic as that which had been well known in the Arctic for decades at this point. The Germans claimed a vaguely defined area 'between' the existing Norwegian and British territories and named it Neuschwabenland, which was recognized by no one.

There was a persistent myth already alive during the Hitler era and continuing through the Cold War that the Germans had established bases in Antarctica during the Second World War. After Germany surrendered, this mythically tinged conspiracy was expanded to include stories of how Hitler was secretly transported from his bunker in Berlin via Patagonia to just such a base, an ice cave in Dronning Maud Land, his new 'Berchtesgaden' retreat, where he could live on after the fall of the Third Reich. As recently as 2022, one of the cornerstones of this conspiracy, by Argentine exile Ladislas Szabo, was republished as *Nazi Antarctic Exploration: Hitler's Escape to South America and Secret Nazi Bases in Antarctica,* the 1947 original in French having been entitled *Je sais que Hitler est vivant* ('I Know that Hitler is Alive'). The myth combines elements from the Nazi imagination with some real elements from the history of polar research in the decades around the middle of the twentieth century. Elements that were later supplemented with aspects of America's Camp Century base high up on the Greenland ice, planned to be surrounded by missile silos aimed at the Soviet Union and with its own nuclear power plant. That base was also real, but like the myths, it buckled under its own weight after only a few years; the ice was more plastic than engineers had anticipated. In recent years, concerns have been voiced that the ruins of the city underneath will appear as the ice melts away, although that's very unlikely during this century. Climate change, which was also unknown to the Americans working on the construction of Camp Century, makes no allowance for geopolitical needs.

The cult of pure, snow-white heights was very real, and interwar Germany did not have a monopoly on it. At about the same time, it existed – in less extreme forms – both as a national ideology and in various subcultures, not just in the Nordic countries but all around the world. Argentina's glacier landscape, crowned by hundreds of thousands of square kilometres of sparkling snow, has gained increasing status and, in recent years, political support, in the form of both the magnificent glacier atlas compiled by the Argentine Institute of Snow, Ice and Environmental Research and an expanded nature-conservation law protecting the cryosphere. The explorations of high-altitude snow and ice were tsarist prestige projects which continued into the Soviet era. One of the crown jewels was the Fedchenko Glacier in Pamir, named in 1878 after the Russian naturalist Alexei Fedchenko, but renamed the Vanch-Yakh Glacier in 2023 as part of Tajikistan's de-Russification policy. It is one of the largest glaciers outside of the polar regions, at around seventy kilometres long and with a total drop of 3,400 metres, and extends down to the border with Kyrgyzstan. Among other things, it irrigates the Amu-Darya, the river that supplied the now dried-up Aral Sea with water. Remote and inaccessible, the glacier was long unknown, but a German–Russian expedition in 1928 and a research station established during the Stalin era brought it to the attention of the public, and its spectacular size and shape gave it an iconic role in the Soviet Union.

The British Empire turned the snows of both India and Pakistan, along with what Ernest Hemingway referred to in his 1936 short story of the same name as the 'snows of Kilimanjaro', into colonial cult objects. In the Himalayas, the process spanned an entire century. In 1965, the British surveyor of India, George Everest, reluctantly gave his name to the world's highest mountain (although it was a certain Radhanath Sikdar who calculated its position and height). In 1953, Edmund Hillary, a New Zealander, had the honour of being the first to reach the summit (although he was there in the company of Sherpa Tenzing Norgay). That's another characteristic of snow – being the arena of achievements and extreme feats which are often themselves political.

The same legitimizing role, at once innocent and entrenching, could be assigned to whiteness in the absolute antipode of Europe, Australia, which is another snow country. Australia already had its own inaccessible, intimidating 'red inland', the vast desert areas of red soil and blistering heat that hadn't been crossed by Europeans until 1861. Australian geologist Douglas Mawson managed to return alive from an expedition to the Antarctic in 1911–14, whetting Australia's appetite for incorporating a 'white' part into its identity. As historian Brigid Hains has shown in her book *The Ice and the Inland*, the country became a nation that holds these deep contrasts: one, a continent of snow; another, essentially of desert. Two very different figures – at least on a symbolic level – embodied this contrast. One, John Flynn, became the missionary of the 'hinterland' and founded the legendary Flying Doctor Service in 1932. Douglas Mawson, meanwhile, was a hero of the ice, and leader of the legendary Antarctic expedition in which he alone barely survived a long sleigh journey, while his two companions perished. Mawson's efforts can be tied to the fate of the British Empire, which Australia left in 1901. Where Scott had just lost the battle for the South Pole to Roald Amundsen, in so doing failing to symbolically win Antarctica for the Empire, Mawson could instead triumph as the representative of a new nation seeking its own place in the world.

So, snow becomes an unexpected component in the geopolitical identity of a nation-continent largely located in the tropics. But there's snow within the territorial boundaries of Australia and skiing has been practised there since the nineteenth century, inspired by Norway. Norwegian ski historian

Jakob Vaage wrote in the 1970s about Norwegian prospectors introducing skis to California during the gold rush that began in 1849. Some of those prospectors later travelled on to Australia, attracted by the news of gold deposits in Kiandra, a mountainous area in New South Wales. Winters there were still severe, with a long snow season in July and August. Skiing was soon taken up by the local population, and in the 1860s ski carnivals began in Kiandra and have been a recurring tradition ever since.

Skiing first took place in Australia in the southern winter of 1861. Only weeks may have separated this event from that of 15 September of the same year, when John King, the sole survivor of the Victorian Exploring Expedition that had crossed the continent from south to north then back to south again, was located by a rescue expedition. In fact, King had already been rescued a month earlier by members of the Yandruwandha people, who gave him food and water. The expedition is common knowledge in Australia, as familiar there as Gustav Vasa's sixteenth-century ski adventures in the Dalecarlia province Dalarna are in Sweden. Fewer people know about the ski races in Kiandra. On one occasion, the Alpine Ski World Cup was held in Australia, in Thredbo, an area in the Snowy Mountains where the country's steepest slopes are to be found. New Zealand has also hosted the World Cup, in Mount Hutt on the South Island, not far from Christchurch. In 2020, a Jewish organization called for the removal of a plaque commemorating one of the ski area's founders, Willi Huber. Huber, it turned out, had been a member of the Waffen-SS, which was originally Hitler's armed bodyguard protection and later a much-feared elite unit responsible for widespread war crimes. No place is free from connections to other places.

Speed and Harmony

*

IT COULD BE SO DIFFERENT.

To me, snow is the embodiment of a better state of things. Snow carries inside it a long-standing sense of morality that we don't seem to be able to live up to. This is the wise snow I encountered as a child in the works of Tove Jansson and Elsa Beskow. I think it is the awakening of my conscience. We read the books aloud before we go to sleep, especially in winter.

These moments are the most important of my childhood. It's mostly my mother who reads, but my dad does, too – when he's not out in the woods hunting moose, birds or hares, or selling Christmas hampers for charity. Before I can read myself, I have to fill in the missing rhymes.

I suspect that what happens in Tove Jansson's book *Moominvalley in November* is also what occurs within myself in late October, when the leaves have fallen and the snow arrives to make the world new again, as in the Alf Prøysen song: 'You will have a day tomorrow that stands pure and unused / With blank sheets and coloured pencils / You can correct all the mistakes from yesterday'. No political platform is more beautiful than these lines. Snow is the Nordic word for hope and forgiveness. We're capable of so much more than what we actually achieve.

The interwar period can be seen as a period of appropriation. Snow was given a role in the rapidly emerging realm of cultural and scientific modernity. The political winds might change, but alpinism – which can also be seen as an ideology in itself – is part of the modernization of Alpine environments. The climber comes along as a cousin of the scientist, but at the same time makes the snow-clad heights part of a conquest project, which at lower altitudes has been about extracting natural resources. Peter Hansen's book on the history of mountaineering is aptly titled *The Summits of Modern Man* (2013). It's modern man doing the climbing. One man. The higher the mountain, the colder and the more difficult it is. So, snow becomes associated with the most difficult level, where the individual – and modernity – are put to the test, through the solo mountaineer and through the technologies that modern societies can employ to access the attractions of nature. According to this interpretation, snow literally stands at the limits of modernity. Snow fascinates us, but it is not allowed to dominate us. Humanity has to be dominant. Through the means of film, tourism and, by extension, warfare, the morality of snow becomes neutralized.

What, then, is the significance of snow? I've already hinted that the answer to this question is partly political and, possibly, most of all historical. The answer depends on what you're looking for in the snowy landscape. On the one hand, mastering the nature of snow is a prerequisite for using it in certain ways. Alpine tourism would be inconceivable without extensive 'terraforming'. That is to say, a reconfiguration of the landscape to include cable cars, ski lifts, roads, hotels, restaurants and now snow cannons to artificially extend the seasons nature has provided. Historian Andrew Denning has investigated how this transitional zone between culture and nature has taken shape. He gives particular weight to the landscape of skiing, in which two ideals, harmony and speed, come together. Speed is one of the defining characteristics of modernity and the sign of humans' ability to transcend the given conditions, to do with technology and (fossil-fuel) energy what ought to be impossible.

At first, the speed of snow isn't that important. In Norway, a little ahead of the curve, 'skiing' became an umbrella term for a range of genres and styles, the practitioners of which were ideally expected to master them all: cross-country skiing, ski jumping and downhill skiing. Early ski clubs were founded in Kristiania and elsewhere in the 1870s, and the sport spread rapidly. Towards the end of the century, Alpine countries followed the Norwegian example, and a large number of ski clubs were formed in Germany, Austria and Switzerland, often as part of existing Alpine clubs. The fact that the mountains were higher and steeper in the Alps led to a debate on the appropriate techniques and equipment that should be used.

Wilhelm Paulcke, a German ski pioneer, led the 'Nordic' faction in favour of transposing Norwegian ideals to the Alps. An 'Alpine' school was founded by Mathias Zdarsky, a Czech artist, gymnast and teacher, when he moved to Lilienfeld in eastern Austria in 1889. Here the verticality of the terrain combined with the abundance of snow made him see anyone serious about their fitness need only focus on skiing. Zdarsky wanted shorter skis for turning on steep terrain with just one pole, which would act as a rudder on the slopes. He experimented with making bindings himself and made two hundred prototypes, patenting one. With his new equipment, Zdarsky was now able to climb steep mountainsides and even get back down on his skis without having to remove them. His new style turned into a business opportunity. Zdarsky teamed up with a local businessman, who helped him develop and sell skis and bindings as a single package. Wilhelm Paulcke, meanwhile, formed a relationship with a competing firm known as Fischer.

Alpine club newspapers and other emerging skiing literature featured hot debates. Both Paulcke and Zdarsky were eloquent ambassadors for their 'schools', which gradually developed into different outdoor styles. Paulcke prioritized cross-country and versatility, while Zdarsky's Lilienfelder technique, as it was called, became more 'Alpine' and performance-oriented. But the contradictions were never fully resolved because, before ski lifts were introduced, all skiers did similar kinds of skiing: touring, preferably in groups, perhaps with some summit climbs. Skiing was very much a way of getting together, a way of using the snow and the terrain socially, and a new way of experiencing nature, mostly for more affluent people. In skiing, unlike many other sports, women and men could participate on more or less equal terms, which was a primary theme in the handbooks. This was especially the case for downhill skiing, where the upright stance and easy, swerving flight downwards indicated modernity and elegance, in comparison to cross-country skiing with its rough arm movements, perceived by the debaters of the time as primitive and 'animal-like', and particularly inappropriate for ladies.

Naturally, Zdarsky and Paulcke each published their own

Anleitung, or instruction manual, which expanded upon their own snow and ski ideologies, along with practical advice. Zdarsky's was published in eighteen editions over the span of thirty years. Characteristic of the development was that the first edition, from 1896, was about *Ski*lauf-*Technik*, while the word in the title of the fourth edition, in 1908, became *Ski*fahr-*Technik*, the latter referring specifically to downhill skiing. So, from 'ski-running' to 'going skiing'. When ski lifts began to be introduced, in the 1930s, these conditions changed immediately. The entire sport was now completely divided, with the landscape industrialized to suit the different aspects of the sport and skiing styles, and in so doing accommodating the rapidly growing number of tourists in the interwar period.

Skiing, according to Andrew Denning – who is thinking mainly of downhill skiing – also requires harmony. Together, speed and harmony created balance, which was necessary for movement to be possible at all. It was this balance, or tension, that made skiing meaningful. The same idea could apply to the snow landscape for tourism, writ large. It presupposed a certain level of nature preservation: living forests, appealing views; that there was natural snow to move on, snow that could endlessly cover the landscape. Balance also required technologies for vertical mobility. Here, too, speed was a factor: lifts, cable cars, departing frequently enough that *skiing* didn't turn into *queuing*. Unless there's a reasonable balance between harmony and speed in such a total snowscape, tourism loses its meaning. In completely 'industrialized nature', as the American technology historian Paul Josephson has called it, without seasonality or natural unpredictability, nature tourism isn't really possible.

Tourism is undeniably a form of consuming the landscape, but it is also, paradoxically, the opposite – the production of a landscape. But this is far from always being the landscape that tourists actually want. Snow tourism, which contributes to the disappearance of snow, undermines its own existence. In the field of cross-country skiing, similar tendencies have emerged, albeit more recently, and sometimes in powerful combinations of different forms of snow sports. Hafjell, Trysil and Sjusjøen are three Norwegian snow resorts where the scale of the facilities – with huge lodge complexes, roads, shops, hotels – now resembles the spatial and functional separation that cities have. Hafjell looks like a landscape-sized workout facility, where lifts, slopes and hard-packed track systems are the equivalent of indoor gym equipment and machines. This kind of snow holds no surprises. It's mute.

It's true that the surface that people move on in these manufactured landscapes is technically snow, however it comes into being. Snow is the right word for it. But it's snow lacking many of the qualities that created the sense of vertigo and magic that constitutes our modern experience of snow. It can be useful to see these tendencies within a greater context, since they belong to the same movement. We've encountered them earlier, in the form of opposing snow ideologies: the play of Hannes Schneider skiing versus the conquest of Leni Riefenstahl's. And cross-country skier Paulcke – he was a great admirer of Nansen, by the way – versus the athleticism and speed of skiing demonstrated by Zdarsky. For people focused on mastery and achievement, the industrialization of nature is the way to go. As is, accordingly, the industrialization of snow, with optimal ski tracks of a predetermined depth, racing tracks with a prescribed amount of vertical metres per kilometre and hard poles. Anything that makes competition on snow unambiguous and makes achievement comparable.

This is well captured by the concept of 'sportification', coined in the late 1970s by American historian Allen Guttmann in his book *From Ritual to Record*. Competitive sports and games played between people may have always existed, and new ones are constantly being created. Those that are taken one step further, to become the kind of sporting events that keep records and have leagues, all go through a process of universalization. Consistent rules, standard equipment, simple criteria for what constitutes winning. Usually they also have made extensive commercial adaptations in service of the entertainment value of the sport.

For the sake of consistency and volume, snow sports launched their own World Cups. The first for Alpine skiing took place in 1967, for men only. A biathlon (a combination of skiing and rifle shooting) was launched for men in 1977 and for women in 1982. Cross-country started in 1982. These were all brought about by Swiss journalist Serge Lang, a writer for the Parisian newspaper *L'Équipe* who understood what the media needed, especially with the expansion into television. Lang wrote in a retrospective, 'a new medium had emerged. The growth of television highlighted the need for a change in the format of skiing competitions.' Having different sports meant that the audiences for those sports all across Europe could be spread out over the European population's free time during the weekend, scheduled with minute precision to optimize viewing and therefore an event's market value. Athletes found themselves acting as magnets for the media and, in time, their advertisers.

And so the formats of the races needed to be adapted. There were fewer long races, while the number of shorter ones increased. The biathlon flourished as an arena sport. Cross-country skiing's transformation into shorter tracks, more sprints and greater drama generated more controversy. The 50-kilometre race in Holmenkollen once comprised two 25-kilometre laps; the audience got to watch the start, a single circuit, and then, after three or four hours, the finish. Today, it's a series of six laps in a more compact space, one after the other, so that virtually the entire race can be seen by the television cameras, and maybe 50 per cent of it by those who have good seats in the stadium. It's over within two hours. Group starts and sprint prizes were introduced.

The transformation gave rise to what were termed 'amoral' victories by sprinters, and to pundits like Norwegian star cross-country skier Petter Northug. There was talk of a spiritual decline of cross-country skiing. Northug became the archetype of 'sportification', fully adapted to the increased media attention. We've got used to it now. The protests coming from traditionalists with conservative values had their merit, but are rare today. This is true even in Norway, where the conservative view persisted for a long time, championed by ponderous media experts and former elite skiers who, in fits of self-deprecating nostalgia, get together online to relive classic fifty-kilometre races on YouTube. Cross-country skiing is said to be the sport that has improved its performance times the most, percentage-wise. This isn't because the snow itself has changed in any significant way. It's because snow has been industrialized.

And mediatized. Households have undergone their own sportification through TV programmes focusing on winter sports. British sociologist Michael Billig points out that sport is one of the really important parts of what he called 'banal nationalism'. Banal in the sense of being fairly harmless, but not necessarily unimportant. And certainly not unengaging. One sociologist who has followed in Billig's footsteps is Norwegian historian Rune Slagstad. In his major work on the history of ideas, (Sporten), from 2008 – yes, there are brackets in the title and a whopping 849 pages between the covers – he was able to show, using Norway as his main example, that sport is a social force of unprecedented dimensions, and any-thing *but* banal. In Slagstad's work, the ideals of sport, of which there are many and they are varied, have both aesthetic and deeply political resonance. Hiking, and for that matter cross-country skiing, tend towards the sensual and aesthetic, and as such can be linked to an expanded, life-affirming under-standing of the citizenship of the welfare state. He contrasts it with the older military, physically rigid idea of sport that has lived on in 'disciplines' such as gymnastics and alpinism.

The twentieth century saw the creation of what Slagstad calls 'cathedrals of the moment', a reference to the other-worldly yet aesthetic nature of sport's defining moments. He, in turn, refers to philosopher Guy Debord's 1967 Situationist classic *La société du spectacle* ('Society of the Spectacle'), best known for inspiring the May Day revolt in 1968. The Tour de France and major football matches are spectacles, sights of the unhaltable present. They are reminiscent of the delight-fully fleeting moments in theatre, opera and dance, but, like revolt, they're of an unpredictable nature, which means that they have a freer relationship to the formal requirements of aesthetics.

The cross-country World Cup competitions, dating back

to the 1980s, can be perceived as a kind of spiritual head-quarters, a central place of televised or streamed connection between all these snow-cathedral moments. From Davos in Switzerland to Falun in Sweden to Otepää in Estonia, they get filtered through a jumble of channels and, yes, media outlets, and offered to their many market segments. An ingenious amalgamation of different social spheres: sports, market forces, faith and art – which in other circumstances would adhere to quite different 'rationalities', to use German sociologist Max Weber's term. Here they are subordinated, roughly but not uneasily, under those of media capitalism. In the background, of course, is the global transformation of the media industry and complex bidding competition for rights and advertising revenue, with national monopolies weakening as new giants emerge and streaming services seep into every corner of our days and our reality.

Maybe you could contend that it's the warmth of the vicarious community, that group of pleasant, chummy participants, that draws the viewer to these mediated communities, in contrast to their own mainly solitary experience. Solitary, because the consumption of this type of entertainment has become increasingly solipsistic. Indeed, our entire society has become more individualized. In the linear media world of earlier times, we participated in the moment, and the media enthralled entire families and workplaces gathered in front of radios and bulky television screens. Now we're individual hunters on our savannah of slender screens, with earbuds in our ears to close even this opening to the outside world. We seek access to the realm of snow by proxy.

Death in the Snow

＊

FOR THE MOST PART, I NEVER THOUGHT about how my own life with snow had begun there in the drifts in the schoolyard. I imagined I was above such things. Until I read about Unn, the girl in Tarjei Vesaas's *The Ice Palace*. I always knew, of course, that people got trapped under the snow and could even die there.

As a boy, in the mid-sixties, I'd read in some magazine about a 22-year-old grouse-hunter called Evert Stenmark, from the town of Umasjö in the region of Tärnafjällen. During a hunting trip in January 1955, he'd got caught in an avalanche and lay trapped there, unable to move except for his left arm, down there in the snow, at a depth of two metres. Because Stenmark had said that he'd be out hunting for some time, it took more than a week before they started missing him down in the village. He was discovered after eight days in the snow, thanks to a nearly miraculous detail. In one of his pockets he happened to have two red movie tickets from a visit to Stockholm. He attached the tickets to the end of a ptarmigan snare made of birch twigs along with the red top to a herring tin, and successfully pushed the twigs all the way out to the surface. The person who ultimately saw this crumpled-up distress flag was Evert's younger brother Kjell.

Evert survived. He could just as easily have been found dead, when the snow melted.

Snow could be dangerous. It would come back to me sometimes. One of those times was during my military service, at Ammarnäs in the Vindelfjällen Nature Reserve, during March in a year with very deep snow, the all-terrain vehicles from Hägglunds' industries tumbling like kittens down the steep hills. Those of us doing our military service got the assignment of digging a snow bivouac for the night. The location had been determined beforehand – the leeward side of a high, steep slope at the upper reaches of the Vindel River, where winter winds had packed the snow to a depth of several metres.

Bivouacking was a way of saving lives during a snowstorm. Useful knowledge. Within a bivouac the temperature would soon reach 0°C, regardless of how cold it was outside. A film of frozen meltwater would form quickly on the inside of the bivouac – the result of a balancing point between our collective body heat and the cold of the March night around us – both reducing dripping and stabilizing the roof. To get rid of the cold air, we were to construct the bivouac on two levels: a larger shelf, where we'd lie in our sleeping bags, and a lower level, where the cold air could sink and escape through an opening. We were told that snow was a great building material, being malleable and stable. You could check the oxygen level with a candle when necessary: the flame had to be burning brightly.

It all sounded sensible. I dug and constructed the bivouac with the same all-consuming energy I'd had as a ten-year-old, and afterwards I was one of the most eager to crawl around in the snow in my white camouflage suit. The bivouac was inspected by the officers in charge and found to live up to all the requirements. It wasn't uncomfortable to lie in, but I couldn't stop thinking about what would happen if the roof gave way. Everything I knew about snow told me that it wouldn't give way, not until May. The melting had stopped, thanks to assistance from the membrane of ice in the roof. My wide-open eyes stared up at the shiny membrane there on the ceiling. It was definitely there. I could see it with my own eyes. But with each moment that I was left to my own imagination, the more my faith sank in that which I thought I knew. As if the distrust within me, which I doubted in itself, was in the end stronger than my belief in reason, which I so wanted to trust. By the time the grey light of dawn penetrated the snow walls, I hadn't had many minutes of sleep.

But still I was in some way proud of my experience. Much later in life, at a safe distance in time and space from the bivouac, I could detail with great conviction what a wonderful lifesaving device it could be. 'There was no wind and it was o°,' is what I'd usually say. 'Really cosy. You can have a candle burning there on a shelf so you know you have enough oxygen.'

In Rondane we dug a shelter for ourselves, too. Not because we had to. We had tents with us. It was sunny, warm, windless, and well into April. But the main mountain station was fully occupied and we had to sleep somewhere. And we knew how to construct one. The snow was looser here and the drift was smaller, but it was okay. It was a good bivouac, and the night was quiet. But it still wasn't for me. At this point I was over twenty-five years old and I knew this wasn't going to change. The fate assigned to me had made me a creature that thrived above the snow, not underneath it.

It was shortly after this experience that I read Vesaas's *The Ice Palace*, pausing when I reached the face in the ice. All my life I have wondered what the worst way to die would be, and concluded that it had to be falling through the snow covering of a glacier crevasse and getting stuck deep down in the palace. And then realizing, fully conscious, that death is going to take a long time. That I'll not be able to shorten my suffering. That I'll have nowhere to go. And that there will be no birch twigs to hand.

On the Kola Peninsula in north-west Russia, a city in the Khibiny Mountains named Kirovsk has a street bearing the name 'Pronchenko'. This is the name of a Soviet mining engineer who tragically died in an avalanche on 5 December 1935 after trying to warn others. Grigorii Pronchenko was a snow hero and deserved to be remembered.

The city that decided to bestow this honour had only received its own name less than a year earlier, in honour of the Bolshevik revolutionary Sergei Kirov, who had recently been murdered and who was responsible for planning the Soviet Union's Arctic mining communities. It had previously been called Khibinogorsk, after the mountain range. The largely treeless mountains, with a maximum height of around 1,200 metres, rise dramatically around the mining community.

Pronchenko had already reconnoitred the mine when he was a young Bolshevik pioneer, and had studied the avalanche conditions in the area, which were known to be troublesome. The Kola Peninsula, north of the Arctic Circle, is subject to many winter storms and the moist winds from the Arctic Ocean mean that there can be significant levels of precipitation. The average number of snow days per year at the time was 220, or more than seven months. A degree of caution seemed warranted.

The administration in Moscow took another position. The kingdom of snow in the north should be, without hesitation, conquered for their new society, which was to be built with Soviet technology and high aspirations for the future. A special snow commission was set up under the Soviet Academy of Sciences, headed by a leading polar explorer, Alexander Fersman. The point of the commission wasn't to delay the northward expansion with Fersman's research, but to control and subdue the snow, in the same way that the Soviet state systematically tried to control all of nature: soil, forest, steppe, rivers, lakes, taiga, tundra and ice.

Pronchenko agreed wholeheartedly. He contributed his own chapter to the book *The Bolsheviks Conquer the Tundra* (1932), which described with eloquent enthusiasm the progress being made. 'Light' was going to be brought to the 'darkness of the tundra', and the first in line to do so were the snow engineers. There were warning signs, but these were intentionally ignored. The indigenous East Sámi people, who had lived in the area for centuries along with their reindeer, were known to avoid travelling on the mountainsides in winter. They knew it was dangerous. Even the initial Soviet surveys confirmed the image of an unusually avalanche-rich area around the mines. In a normal winter there were ten to twenty avalanches. When more systematic investigations in the Khibiny area began in 1933, it turned out that avalanches with a volume of at least 200 cubic metres occurred even more often than previously thought. A summary of developments up to 1938 showed three hundred avalanches, which was more than fifty per year. The warmer climate in the North Atlantic and around the Arctic Ocean during the interwar years, the so-called 'Arctic warming', may have increased the frequency of avalanches somewhat during this particular period.

It was typical of the Soviet state not to let these kinds of circumstances interfere with planning. The first five-year plan in 1928 had identified the Khibiny mountain mines as a major supply of metals and minerals to Russian industry, and there was no backtracking on that position. The mines were particularly important: they contained apatite, a component in fertilizer that was a key element in the Soviet drive to nationalize and expand their agriculture. Thousands of people, mostly young men, were hurriedly recruited to the new mining communities. Pronchenko himself was one of them. Many were assigned housing in the new Soviet town of Khibinogorsk, a short distance from the mines. Others were housed in simple barrack-like buildings just under the mountain, right next to the mine. Work was carried out outdoors on the mountainside, indoors in various processing plants and underground.

The outdoor places were the dangerous ones. According to environmental historian of Russia Andy Bruno, it was mainly kulaks, formerly wealthy peasants whose lands had been taken, who were allocated housing near the mine. Many kulaks were forcibly exiled to the Arctic mining and industrial projects, where they lived without many of their civil rights, including freedom of movement. The Bolshevik party's vanguards and experts of various kinds, including mining engineers like Pronchenko, were placed in the same at-risk environment. They were to make meteorological observations and carry out geological surveys, tasks that required a constant presence.

It wasn't long before the dreaded accidents did indeed occur. Already, during the initial winters of the 1930s, avalanches had come uncomfortably close to the dwellings. Some of the avalanches had damaged buildings, and a few people had died. But there wasn't a disaster. Then, in December of 1934, a massive avalanche took place that was to claim the lives of 86 people, most of them sleeping in simple dwellings at the foot of the mountain. The death toll may seem modest, but for an avalanche disaster it was very large, one of the biggest ever, on par with the infamous Wellington avalanche in Washington State in the US, on 1 March 1910, which killed 96 people, and the Rogers Pass railroad avalanche in British Columbia just three days later, in which 58 people lost their lives. Compare these figures to the current annual global avalanche death toll, which is usually between 100 and 200 people.

The event was reported in the Russian and international media, prompting the local authorities to set up a special snow monitoring unit. A year later, an avalanche killed Pronchenko while he was on a mission. With each fatal avalanche, the planners' brows grew more furrowed, but mining production remained the priority. In February 1938, there was another huge avalanche. This time 21 lives were lost.

The lack of regard shown for the risk of death in the face of Khibiny massif's avalanches, and the unreasonable prioritization of production plans over considerations of risk and human life, fall into a larger pattern of contempt for the people's suffering that can be tied to the unyielding centralism of the Russian Empire. The roots go deep, back to the reign of the tsars, and by the time Soviet industrialization was in full swing the pattern had become systemic. At the same time, Khibiny was an exception and, in many ways, a turning point. These repeated accidents eventually became unsustainable, not only because of the loss of life, but because they made it impossible to preserve the credibility of the Soviet Union's Arctic strategy. Risks had to be reduced, and the Soviet state mobilized all possible forces to try to prevent what otherwise threatened to become an avalanche crisis in a strategically important region.

Knowledge was the greatest power. There were research projects in snowy regions all over the empire, and Arctic research was of particularly strategic importance. The Soviet Union's major investment in the northern regions mainly focused on the Northern Sea Route, the *Glavsevmorput*. If this sea route along the Russian Arctic coast could be opened to cargo ships, Russian raw material exports would benefit, and the Soviet Union would become more competitive relative to Europe and other parts of the world. Research on ice and snow became a top priority, and by the 1930s the Soviet Union was a leader in most kinds of Arctic research. It began its famous ice floe research in the Arctic Ocean in 1937 with research stations on the moving ice that could support dozens of researchers.

Avalanches had attracted notice before, for example during road construction in the Caucasus in the late nineteenth century, but they hadn't emerged as a major priority until now. Avalanches stood in the way of efficient production of minerals in the same way that the cold and sea ice did, but they risked fewer casualties. Data from the Khibiny Mountains became part of the resources that geographers and naturalists began to use to gain a better theoretical understanding of the physics and dynamics of snow. So, Khibiny became a place that new generations of Soviet avalanche researchers sought out. One of these was Georgy Tushinsky, who soon became a leading expert on snow, establishing operations both with the help of the mining company in the region, Apatit, and the Moscow State University. His book *Laviny* ('Avalanches', 1949) became a standard work. In *Sneg i laviny Chibin* ('Snow and Avalanche in Khibiny', 1967) he collected results from a number of contributing researchers.

The outcome was a greatly improved avalanche warning system and a wide range of preventive measures of the kind that had been in place in Switzerland for decades, with avalanche barriers on the slopes and continuous monitoring of snow depth at strategic locations. These developments in Switzerland were already an example for the world. Indeed, Alexander Fersman's efforts had been inspired by Switzerland. Fersman had been in contact with the eminent crystallographer Paul Niggli, who was a leading organizer behind the development of Swiss avalanche science in the interwar period. On a visit to Davos, Switzerland in June 1936, Fersman learned the importance of preventive blasting and saw the success of avalanche fences and tree curtains that could catch avalanches once they'd begun. Fersman met Niggli and the 'energetic young professor' Henri Bader, and returned home with the latest publications from the Davos research. Above all, he had learned how to set up avalanche monitoring systems.

After a dark period in the 1930s, with a total of 120 deaths, only a small number of fatal avalanches occurred in Khibiny over the next sixty years. The death toll remained a handful per decade and mostly tourists, as the mountains became increasingly popular for growing ski tourism. The extensive and internationally prominent research had borne fruit and a safety mindset had taken hold.

The worst avalanche we know of happened in Peru in May 1970. The town of Yungay and several small villages were flattened beneath the 7,000-metre-high Mount Huascarán, in a combination of avalanches and landslides triggered by an earthquake. Twenty-two thousand people died nearly instantly. That's ten times the number of people who died in Pompeii during the two days of fire, ash, dust and rock spewed by Mount Vesuvius in the year 79 CE, after the preceding earthquake. While the prolonged process in Pompeii allowed the majority of the population to escape with their lives, the inhabitants of the Peruvian mountain town and villages met their death in a few seconds in the middle of the day. Because it was Sunday, many people had been out in the public squares.

Avalanches of the past have wreaked similar destruction. Much of it is likely unknown to us, buried not only in the snow that fell on villages and people, but also due to collective amnesia. The death toll was rarely high, however, because the population was small and the groups living in avalanche areas very few. However, a fascinatingly large number are documented in several select countries, such as France, where avalanches have been recorded since the twelfth century. The one that hit Chèze in the Pyrenees in February 1600 was particularly devastating, killing 107 people and destroying many buildings. And the Nordic example of the two devastating avalanches in Møre and Romsdal in Vestlandet, Norway in 1629. Also in Norway, there was the

Skylstad avalanche in Norang, and the Valset landslide in Bondalseidet, which saw 28 and 27 dead respectively, both in the municipality of Ørsta.

The worst avalanches in Europe, in terms of death toll, were in the early 1900s. During the First World War, tens of thousands of soldiers died in avalanches in the heavily fortified Alps, triggered by a whole host of causes: risky troop movements; poorly chosen bivouac sites; artillery fire that set them off unintentionally, long before bombardment became an established method of preventing avalanches by triggering them in a controlled manner. The death toll is uncertain. There are reports of 40,000 deaths within just a few years, but also of double that number. In December 1916 alone, as many as 10,000 men (certainly at least 2,000) may have died in avalanches on the Italian–Austrian–Hungarian front. At the Marmolada massif, three hundred Austrian soldiers died on 13 December when an avalanche of snow, ice and rocks crushed their living quarters.

During the interwar period, and for a good bit after the Second World War, major avalanches continued to happen throughout the Alps. In January 1951, after a period of heavy and prolonged snowfall, two major weather fronts, one from the south and one from the north, suddenly converged, bringing an additional two metres of snow in just a few days, along with rising temperatures. In Andermatt, Switzerland, six avalanches were triggered within the span of a single hour, destroying restaurants, hotels and shops. Thirteen people died. In total, 649 avalanches were counted in the Eastern Alps during that month alone. In Austria, which was the worst affected, 135 people died and the material damage was enormous, with villages destroyed, houses crushed and entire forests flattened. In Switzerland, 98 people died, most of them in the canton of Graubünden, and hundreds of cattle were also killed.

Just a few years later, the Austrian village of Blons was hit. The villagers were acutely aware of the avalanche danger. But on 11 January 1954, there were two avalanches, one in the morning, one in the evening, that were more powerful than anything they'd experienced before. A third of the village's population of nearly four hundred people was swept away by the avalanche, many of them in their own homes. The death toll eventually reached 57, including two who were never found.

The detail provided by eyewitness accounts is sometimes hard to absorb as the power of even fairly ordinary avalanches can be hard to imagine. 'At Gletsch,' it says, 'a large part of a multi-ton iron bridge was hurled nearly fifty metres *upward*.' Another story tells of 'an Alpine stagecoach thrown across a river, coachman, horses and all, as it approached the Flüela Hospiz in the region of Grison.' There are legions of similar observations. The velocity of a wet-snow avalanche can exceed two hundred kilometres per hour. Wind avalanches, where a vacuum forms in advance of the front and 'sucks' the snow forward, can be even faster.

But the overall trend since the 1950s is that fewer and fewer people are dying in avalanches, especially in relation to the number of people moving about in avalanche-prone terrain. On one level, the reason is simple. Before the advent of snow tourism, fewer people had reason to travel over snow-covered mountainsides. Most avalanches were spontaneous, without any human presence, and went unnoticed. However, when an avalanche hit a village or populated valley, it was all the more devastating because communities didn't plan to avoid avalanches (although they might avoid rebuilding where an avalanche had struck). Old houses, barns, stables, fences, bridges, mills, forges, factories, churches and vehicles were unlikely to withstand the force of avalanches. Meanwhile, individuals could be hit by avalanches as they travelled through the landscape, if they were unlucky enough to trigger one.

By contrast, a mountain village in the shadow of a protruding downwind-hanging snowdrift was exposed to the same risk for much of the snow season, for weeks or months. The same was true if the village was located under a steep slope with a lot of gravel, where the warm early spring sun would shine on the snow higher up the mountain. When the avalanche struck, sooner or later, as it always did – out of reasons of pure probability and through the intervention of gravity and the other laws of nature – the village could be seriously damaged. The death figures, as we have seen, could easily reach double digits. For a long time, these events were deemed natural disasters and were seen as inevitable fate or God's punishment.

But as major infrastructures in industrialized societies moved more and more of their operations into snowy landscapes, the whims of the snow could no longer be tolerated.

All landscapes become industrialized when the space, the terrain, be it formerly wasteland or wilderness, is transformed into productive land for resource extraction, energy production or transportation. Just as the endless coasts of the world's oceans are equipped with lighthouses, which are electrified and automated. It's an acceleration into overdrive of what had been the slow terraforming of the earth that's been going on since humans first appeared. In this industrialization, only certain forms of snow pass through the needle's eye of profitability. Avalanches don't fall into that group. Their kind of power isn't useful or predictable; it's exclusively destructive.

Snow resting on lakes, streams and rivers in not-too-hilly parts of northern Europe, on the other hand, was useful. The pause itself was important – for the snow to rest until the right time came. Snow acts as a brake, a resting point. It's a waiting game. When the snow melted in spring, its positional energy could be used to transport timber downstream to factories and ports. This was necessary in a country without roads. Only with the development of forest roads and large-scale trucks – and abundant imports of cheap fossil fuels – could this last important step in the industrialization of the forest ecosystem be taken. In Sweden, this happened only in the second half of the twentieth century, up until the 1980s, by which time so-called timber rafting had virtually ceased. In the forest itself, the horse had also been dismissed, replaced by forest tractors and harvesters. Forest workers commuted to their workplaces by car and minibus. Skis had become equipment for outdoor activities and slalom. No one lived for months on end in small huts between the spruces with their deep-snow skis propped against the wall. Snow was no longer an asset, but an obstacle.

Most of this kind of industrialization of nature, and certainly the most large-scale of these projects – the power-line corridors, the power-plant dams, the roads and railways, the clearing of forests, the steel grimaces of wind turbines – has come about in the last hundred years. Once it happened, it was hard to reverse, despite the damage to ecological systems, environment and climate. We made gains in terms of economic efficiency, and we were able to increase our prosperity. But we lost a lot. We failed to transform the world into any reasonable form of equity. By all accounts, we're losing the snow itself now, too. Some risks, especially local ones, could be reduced, although dam failures and flooding disasters have been relatively common and continue to happen. Most importantly, we failed to realize in time how the entire production of wealth through its global linkages had consequences that turned out to be planetary, but at the same time would hit people and cultures differently and on a local level. Worst hit are the places where freshwater supplies run out or tourist income is lost.

Nobody wants a return of avalanches, those messengers of death from the seemingly silent snow. But their removal from civilization follows the same overall logic. It's the modernization of snow that has relegated avalanches to the background. Ironically, it took mining companies (as in the case of Khibiny), railroad companies (Washington and British Columbia), passengers, tourists and their hotels and restaurants (as in the Alps) to up-level the safety of shepherds, farmers and agricultural workers. Switzerland, again, is a good example. On 23 December 1919, an unusually large avalanche that hit the village of Davos damaged buildings and killed several people. This fatal avalanche was one of the many across Switzerland that, taken together, justified the focus on avalanche research. The Alpine village with the steep mountains was situated perfectly, eventually becoming a rapidly growing and important destination for winter sports. Davos as a destination, and even Switzerland as a nation, had begun to realize that it had far too much to gain from the tourism industry around snow to put up with the risk to visitors from avalanches.

The cost of avalanche barriers, to protect the town against anything similar to December 1919 happening in the future, was projected to be a million Swiss francs, a dizzying sum in the interwar period. But the Swiss paid it. Another major avalanche occurred in Aletschwald in 1931, killing a Swiss guide and four British tourists. There could be no worse advertisement for tourism. Stone or wooden railroad barriers in avalanche-prone areas were major investments. Greater still were the costs of railway tunnels, which of course were the only way to fully protect roads and railways from snow. The Swiss met the cost without blinking. Without anyone complaining. Trains in Switzerland run like clockwork.

The political context was quite different, and the natural resources of a different sort, but in many ways the logic is similar to that in Khibiny. Modernization cannot tolerate the wrong kind of snow, or snow in the wrong place, in the wrong shape, in the wrong amount, coming at the wrong speed. By the 1920s, snow had come a long way from the crystals on the sleeve of Kepler's coat, posing questions that Descartes couldn't have imagined. Snow is no longer simply the stuff of beauty and physics, but also of ethics and politics.

And of technology. Engineer Pronchenko enters the story of snow with his lofty ideals of bringing in light and conquering the tundra. The Swiss called in engineers to experiment with snow barriers along the mountainsides. Ideas were drawn from military technology (anti-tank obstacles) and from hydraulic engineering (guide arms, stone chests, ironsides, ferrules, dams). All to divert the snow. And if there was still a risk of the snow getting too close for comfort, you could build a snow wall up against the mountainside to distribute the force. In older times, buildings like churches were equipped with a plough-like extension towards the mountainside so that the avalanche literally split when the wall of snow struck it, like the bow of a ship meeting a storm wave on the sea.

In the mid-1930s, once again in Switzerland, preventive detonations against threatening avalanches began, using the method that had previously been a side effect on the Alpine battlefield during the First World War. Most effective proved to be 81-millimetre artillery guns, but even smaller weapons worked. Along the railways, pre-emptive rifle shots could be carried out at critical passages. Gradually, the same method was applied in ski areas. Soon the explosions became a matter of course, part of the acoustic landscape of snow.

A few generations later, the whole thing is portrayed as a farce – or a tragedy. This is precisely the approach taken in Ruben Östlund's 2014 film *Force Majeure*. When the whirring of the ski lifts has fallen silent around the hotels, the sounds of shots in the night – accompanied by the archaic mechanics of the piste machines and the hissing spray of the artificial snow cannons – symbolize both the soulless industrial nature of this commercial snow landscape and the repetitive sameness of the vacationing Swedish family's unimaginative repertoire. In the film's key scene, an avalanche comes down anyway, as impossible to keep at a distance as all other nature. A thundering, wordless messenger of the wilderness that, in a white shroud of death, shocks all the high-paying guests with its intrusive frankness. It's traumatic for the well-meaning, but lost, family father. But perhaps above all a memento for the middle class consuming the planet, if it's even remotely capable of understanding the role it plays in world development. The film's final retreat, on foot, down the serpentine path of the Alpine slope, leaves the question unanswered.

In other words, avalanches become a central component, indeed a preoccupation, with the emerging narrative of snow as not just some petty obstacle, but as a real threat. The first studies of avalanche danger in the Alps appear in the mid-nineteenth century. The English word comes from the Old French *avaler*, 'going down', literally 'towards the valley' (after *val*, 'valley'). *Lawine* is the German word, which has become the one we use in Swedish. It comes from Rhaeto-Romanic, via the medieval Latin *labina*, 'landslide', from the Latin root *labi*, 'slide' or 'fall'. *Lawine* is related to words like labile (prone to fall) and lapse (a slip-up or a break in continuity). It's all about the pull of the earth. But unlike the snow that falls down into the valleys at a modest pace over a long time, the avalanche is sudden, heavy, dramatic and immediately threatening to life and property.

The history of protecting against avalanches is a long one, going back more than a century. Swiss hunter, alpinist and surveyor Johann Coaz took an interest in the issue. Educated at the Forestry Academy in Saxony in the early 1840s and responsible for the forests of Graubünden and St Gallen, he'd become increasingly interested in Switzerland as a nation of forests. Through dendrochronology and the discovery of older trees, he realized that the forests of the past had once been much larger, even there in the Alpine regions. He writes about this in his first book about forests, *Der Wald*, published in 1861. He advocated for more and richer forests, as this would be good for Switzerland.

As his ideas gained acceptance, he rose through the ranks, becoming a trusted voice.

During his many years of mapping the Swiss landscape, Coaz sees the damage that avalanches cause, especially to younger forests, and he thinks that avalanches should be managed, ideally prevented. At the same time, forests had been the old, established barrier against avalanches for hundreds of years. And that had worked. In other words, it had been a win-win situation: if you can protect forests from avalanches, you will at the same time protect the people in the Alpine villages, who can then grow even more forest on their land, which in turn provides protection – and income. But for this to happen, mechanical avalanche barriers had to be strengthened. In earlier times, farmers were already creating barriers in the form of ditches, terraces and walls of stone and earth to divert the advancing snow. But Coaz realized that this wasn't sufficient.

For Coaz, avalanches are 'fixable' snow. He collects his thoughts on this matter in several writings. One of the earliest is *Die Lauinen der Schweizeralpen* ('The Avalanches of the Swiss Alps') from 1881. By this time he is already the national forestry master, and has been experimenting with avalanche barriers in certain places, and thinking about their construction, since the 1860s. But only a few of these come to fruition. He stays engaged in the subject, mobilizing interest and giving advice. He also gathers information through field research on what conditions – weather, vegetation, geology – make certain places more avalanche-prone. And so elementary avalanche research emerges. In 1910, he publishes *Statistik und Verbau der Lawinen in den Schweizeralpen* ('Statistics and Control of Avalanches in the Swiss Alps') and patiently continues acquiring knowledge on the subject. By now, new avalanche barriers (*Verbau*) have been put in place in many of the Alpine valleys, often high up on the slopes to reliably prevent the snow from moving downwards. But there are still far too few of them.

Avalanches have always been a known danger in the Alps. But now an element comes into play that will soon fundamentally change the interest in them. This figure is the snow tourist, first in the shape of winter climbers and cross-country skiers. In the interwar years, more and more downhill skiers and growing crowds of pleasure-seekers started staying in the increasing number of luxury hotels. Travel writers from the UK who supplied the Alpine countries, and also Norway, with the first batches of ski tourists, took as their subjects these areas and people. Books with titles like *The Ski-Runner* (1903) and *Ski-ing* (1910) made avalanche danger a concern for those who read English and were expecting to ski in avalanche-prone terrain. German-language literature also emphasized the danger. Austrian mountaineer Emil Zsigmondy did so in 1885 in *Die Gefahren der Alpen* ('The Dangers of the Alps'). This book continued to be given out by Wilhelm Paulcke, the northern German ski promoter, who was also a geologist and avalanche researcher, and gradually expanded his repertoire as a far-sighted alpinist. Like Coaz, Paulcke was a proponent of greater knowledge, both among the growing number of tourists visiting the Alps and about avalanches in general, through research.

Similar to most of the others who wanted to conquer avalanches, Paulcke believed that research needed to be done out in the field. Science historian Dania Achermann has emphasized the patriotic element in this form of avalanche control. All of Switzerland had to unite to curtail this great danger to the country. Gathering information by mobilizing reporters from among the people was a method of compiling knowledge about avalanches and snow that had been used in Switzerland in the nineteenth century and which Paulcke, who was himself from Germany, eagerly latched on to. His vision was an 'outdoor laboratory' to which everyone had access, in a variant of what has come to be known as 'citizen science' (the term dates from the 1980s). In this way, Alpine clubs, and hiking and climbing societies, also had a function in forging ties between the tourist elites of the cities, the farmers of the smaller Alpine villages and the emerging tourism industry. Ski instructors themselves had a central role to play. Their wide-ranging expertise and coordinating functions became of utmost importance, and they occupied a top position in the approach that was now being put together.

Safety precautions were already in effect in advance of the winter Olympics in Chamonix in 1924. Savoie and Haute-Savoie, where Chamonix is located, are among France's most avalanche-prone regions and had recently experienced several fatalities. Swiss geologist Henri Lagotala attempted to use scientific methods to model the power of an avalanche and its ensuing effects, and various types of avalanche modelling would soon be applied in many locations. Another type of modelling involved calculating avalanche probability, depending on snow conditions, precipitation and wind. These factors combined could be used to calculate the seriousness of the danger – derived from the combination of the hazard and the probability. Scientists soon realized the potential of avalanche forecasting and the concept of avalanche danger, and after the Second World War a four- or five-point quantification scale was introduced for avalanche danger forecasts. The public were warned regularly, first in the Alpine countries, but soon in other parts of the world, too.

Avalanche literature focused mainly on those who were temporary guests in snowy landscapes, and far less so on the inhabitants, who were unlikely to read instructional books on how to live their lives. The literature is practically oriented, often included in ski instruction manuals and advice for mountain tourism, which began to spread in many countries during the first decades of the twentieth century. Most of the authors who analysed avalanches and made careers as military officers also became instructors and advocates for the new sport of skiing. Paulcke himself was one such example. Some were also scientists, though their research style was often spontaneous and descriptive rather than methodological. Personal experience played a significant role, laboratory experiments were rare and theoretical considerations were rarer still.

Similar developments were taking place in other parts of the Alps. The career of Georg Bilgeri is informative. An officer of the Austro-Hungarian Army's Fourth Tyrolean Hunter Regiment, he had little interest in the physics of snow or the natural geography of avalanches. He saw it as his duty to win on the battlefield of saving human lives – whether that meant soldiers' lives or those of ski tourists – and to promote skiing, for which he was already giving lessons at the age of twenty-three. His role in the First World War was as a leader of Alpine soldier training and a writer of instruction manuals for mountain warfare. His main contribution to ski technique, put forward in one of his texts, *Der alpine Skilauf* ('Alpine Skiing') from 1910, was an insistence on using two short poles in difficult mountain terrain. This was called *Zweistocktechnik*, which, in combination with the *Stemmbogen* ('plough turn'), revitalized skiing and made the slopes passable even for those with novice, tourist ski-legs.

Bilgeri was multi-talented, developed life-saving techniques, was an expert in equipment (ropes, bindings, ski boots, ice axes, climbing skins, avalanche probes, backpacks, ski wax . . .) and for several years ran a factory manufacturing military skis. He trained border guards and police officers, and founded and worked for several Alpine ski clubs. According to one account, he instructed and taught over 40,000 skiers in ski technique and safety in his lifetime. His name and reputation travelled far, and he was soon invited to share his knowledge in aspiring ski nations such as Hungary, Turkey and even Sweden, where his opinions about 'Alpine skiing technique' were featured in the Swedish Ski Association's 1926 yearbook.

Bilgeri is an extreme example, but still typical of a prominent form of snow authority who responded to the demands of the market. First and foremost, they were in high demand on the ground. Skiing had become popular, both in cross-country form and as part of a wider winter alpinism that included ski mountaineering. Bilgeri was the champion in this realm, having completed countless summits. Another growing need was the military. The geography of war expanded with motorized vehicles and artillery, from the rigid battlefield formations of previous centuries to more mobile battles, even up among the higher mountains, which were under threat of avalanche. Forces in that space required protection, but they could also use avalanche triggering as a form of combat.

Male hegemony in this literature was essentially total, stemming from the multiple masculine codings of the field: military, engineering, scientific, alpinist, proletarian. The last term refers to forestry work, from which many of the male racers in the Nordic countries were drawn. In the Alpine countries and the Soviet Union, they often came from the mountain police and the military, respectively. The boundaries could otherwise be unclear between the growing number of activities linked to snow: tourism, competitions, surveying, defence activities, avalanche safety and research. The great interest that royals have in ski sports has its roots here. Particularly in the Nordic countries, skiers, with their physical strength and endurance, were like front-line soldiers on which the values of the nation rested, above all national survival. When a Swedish or Norwegian king shook hands with the victor of the 50-kilometre race, their reverence was mutual. That of a subject before their ruler was standard protocol, while that of the king to the skier was not because the latter was a worker, but because he, too, was in fact a king, a ski king, and as such bore a nobility that was ultimately a foundation of true majesty. Through recognition of one another, they signalled their dependence on each other, a symbolic sign to the whole population of the importance of kings.

These two significant movements – snow propaganda and snow security – were in general separate, but they were defined by, and reinforced, each other. On the one hand, there was education about snow for the new mountain tourists on their skis, and the societies and associations they relied on. On the other hand, there was the technical design of avalanche barriers in risk-prone areas and its scientific underpinning in the form of laboratories for snow and avalanche research, with Switzerland at the forefront as a primary incubator of knowledge that spread across the world.

The exclamatory punctuation in Walther Flaig's 1935 title *Avalanche!* is an expression of the matter at hand. A subject that also touches on a myth about snow and avalanches. Flaig posits that a person can trigger an avalanche with their voice ... But is that true? In their 2009 article 'Avalanche triggering by sound: myth and truth', two Swiss researchers, Benjamin Reuter and Jürg Schweizer, examine Flaig's thesis. Reuter and Schweizer carefully reviewed the research on acoustic influences on snow stability and established that sound can indeed trigger avalanches. Sound waves comprise a wave energy that exerts pressure on the surface of the earth. Things like jets taking off, sonic booms and helicopters landing can all exert that level of pressure. But a human being alone can't make that level of sound. A human scream produces a pressure of two pascals; a jet plane taking off produces pressure of twenty pascals; a sonic boom produces two hundred – a hundred times stronger than the human throat, that weak instrument with which nature has equipped us to summon help and save lives, our own and others.

I've held my breath in precarious situations myself – I think it's a reflex. *Whisper. Hold your breath inside your chest. Make yourself as light as possible, even though you can't.* I am now cured of this minor preconception. But if I stood there on the loose snow again ... would I really dare to scream?

Gerald Seligman, the British geographer and glaciologist, wrote about quiet: 'Keep perfectly silent.' But he advocated silence for another reason: the avalanche makes itself known through sound, even from far away. We have to be silent in order to listen. Skiers who sense danger, or even just risk, should also loosen their ski bindings and their backpacks. Anything to prepare for the avalanche that might be coming. Skis are the biggest burden. Deep down under the snow, still-attached skis are as deadly a restraint as they are angelic wings that allow us to soar on top of it. Seligman said he knew of one case where a skier fell into deep fluffy snow with his skis on and suffocated to death.

Whether Seligman actually had basis for what he claimed is hard to say. Whatever the case, we know now that avalanches don't have to make themselves known due to sound. Death can whisper. For all his gentlemanliness, Seligman was always that way as a writer: assuring the reader of the truth, but sometimes omitting the supporting evidence. The truth is difficult; it doesn't always have clean edges. The American naturalist Ruth Kirk testifies in her fine 1977 book *Snow* that the craters the wind sometimes creates around a tree can operate like a trap, and anyone not paying attention, and who falls into one of these while out skiing, can have a hard time getting out.

Many people have died in the snow, but the snow itself is seldom the reason for their deaths; rather, there may be something else that they have in common. Franklin, Andrée, Scott, Wegener: they all fell victim to their own overconfidence – in new technology, in the superiority of the West, and in their own abilities, which in reality were modest. Then there are all the unfortunate off-piste skiers, who now comprise the majority of the world's deaths in snow during a typical year on Earth.

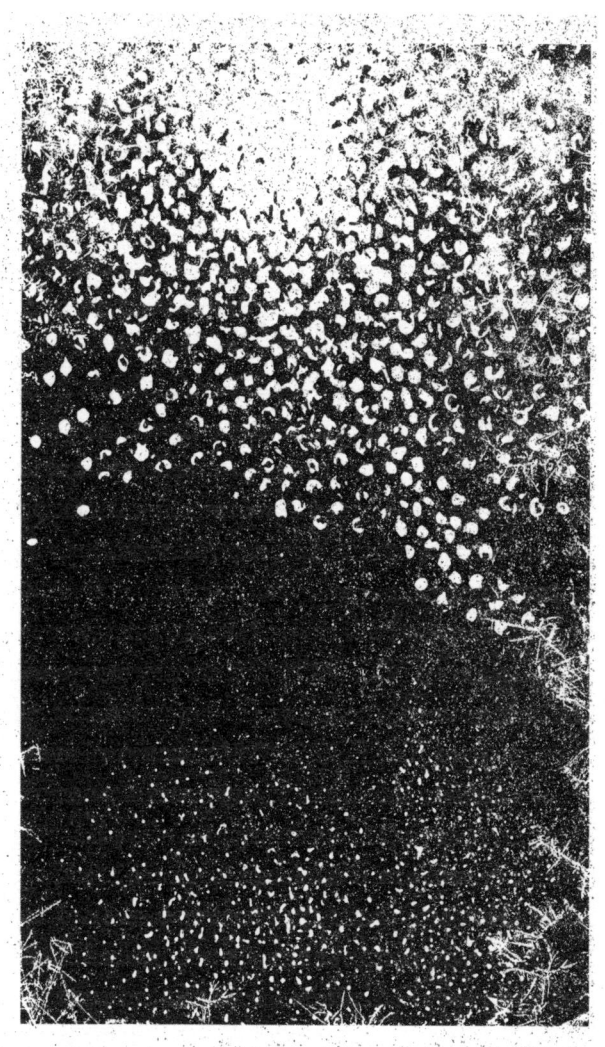

Metamorphoses

*

IN AUGUST 2014 I FIND MYSELF IN DAVOS. I want to visit the real-world version of Berghof, the sanatorium where the author Thomas Mann's novel *Der Zauberberg* ('*The Magic Mountain*') is set. Nowadays it's a hotel and museum. It feels odd to find yourself in a world which is, for me, primarily literary. I'm looking out over the beautiful restaurant, with its two floors, just as it is in the novel. Those of us on the guided tour get to go into a typical patient's room and to see the balcony where the people with lung ailments would lie for hours between their lavish meals.

It's an interesting visit, but it confirms an old conviction of mine that literary places are usually diminished when they step into reality. It's primarily something to do with scale. In a book, a lot can happen in a small space. Prolonged periods of time can move quickly. A single moment can be extended into an eternity. Literary worlds defy time and space. This meeting between objective and subjective time is one of this novel's themes.

At the other end of Davos, I visit the Institute for Snow and Avalanche Research, which was established during the Second World War. This is a world of facts. I take in Swiss avalanche statistics and see amazing maps showing which places were hit, when and how severely. Snow depth, storms, snowfall. Outside the institute, a small stream trickles by.

I really perceived those worlds as being separate. It snows a lot there in the mountains, and *The Magic Mountain*'s protagonist, engineer Hans Castorp, goes skiing in the book – which was published in 1924, the year of the Chamonix Olympics – which leads to Mann's Nobel Prize in Literature. Skiing seems, for a brief moment, to be the most modern thing there is.

I had started to wonder if there might be a deeper connection between the sanatorium and the institute. My intuition told me that there ought to be something about the place, something about the snow, that allowed Davos Dorf (where the sanatorium is) and Davos Platz (the site of the institute) to speak to each other. They ought to, *had to*, speak to one another. I sensed this without seeing it clearly. The way you can make out a human face behind a wall of ice.

Hans Castorp comes to the sanatorium in Davos to visit a friend. A short visit, he thinks, but he is to spend seven years on the mountain. Much of this time in Thomas Mann's *Der Zauberberg*, it snows. During Castorp's second winter, it snows so heavily for several weeks that the usual Alpine sun is never able to show itself. Castorp is seized by an impulse. He must take the snow seriously. The snow, he has gradually begun to think, is so much more than a banal winter covering draped around the salons and balconies of the luxury hospital. But what is it?

He makes his way down to the village, Davos Dorf, where he buys a pair of 'de luxe' skis, as he calls them. Castorp is not a sports nut and has no experience of skiing. Nor has he any knowledge of the great mountain that towers above the sanatorium. After a few days' practice, however, he's learned what it takes, because 'one quickly gets readiness in an art where strong desire comes into play'. What does he have such a strong desire for? To uncover the world, as it turns out. He senses from the snow-capped mountain a wordless answer to a deep riddle.

He sets out, in thin clothing, reinforced only by a simple camel-hair waistcoat. In the deep snow, Castorp discovers another world. Up on the mountain, the snow glistens – like sparkling diamonds. 'Always these diamonds!' Castorp finds the snow 'fairylike and comic, an infantine fantasy . . . the droll, dwarf-like, crouching disguise all ordinary objects wore made of the scene a landscape in gnome-land, an illustration for a fairytale'. He likens the snow to the seashore. Snow and sand are both loose, airy, dazzling white materials. Alien from the deadly corrosive stains found inside his own lungs. Snow and sand also have in common that they are made of crystals.

In the snow, Castorp meets a deeper silence than anywhere else: 'listening to the primeval hush, the deathlike silence of these winter fastnesses'. He overcomes his own fear. Skiing is about facing life's challenges head-on. Castorp discovers, to his astonishment, that he is braver than he thought. He feels the fear of death, but masters it out of a sense of 'kinship'. With what? We are told that 'in his narrow, hypercivilized breast, Hans Castorp cherished a feeling of kinship with the elements'.

How should we understand this idea? What are the 'elements'? Thomas Mann has Castorp experiencing an emerging understanding of the tourists, the 'silly people on their little sleds'. The snow thus becomes a fate and a destiny that forces him to try out the experience that's about the real content of life. Alone in the silence, he can feel his own heart-beat, clearer than ever. 'A naive reverence filled him for this organ of his, for the pulsating human heart, up here alone in the icy void with its question and its riddle.'

A storm is brewing, which he stubbornly dismisses as a frivolity. Instead, he purposefully revels in his conviction that he is in the presence of death, but still cannot let go of the idea that he must not take too many risks. He risks collapsing and freezing in the snow. Despondency and anxiety swirl around inside him with the same intensity as the white flakes rotating around his swaying body in the storm. Eventually, he spots the same barn he had passed a short while ago. He takes shelter under the eaves and drinks, perhaps carelessly, from a bottle of port he has brought.

He falls asleep against the wall of the barn and soon starts dreaming and hallucinating. Visions of Paradise, of sweet moments around the Mediterranean. Death visions of human cruelty. He sinks into the snow and in its pleasant warmth he is embraced by the depths of the dream. It becomes an embrace with death. In this moment of half-awake vertigo, he realizes the deeper truth of the sanatorium stay, which is that death and life are related. Just as illness and health are not opposites. Nor are spirit and nature.

As much as Castorp has fallen in love with, indeed united with, the snow, he has come to realize that the whole of life that the snow represents is in contrast to the hexagon that the experts say the snow is made of. Over and over again Castorp displays a profound contempt for all the crystalline and regularly hexagonal things that science pushes on people.

This is a central point in Thomas Mann's novel: the

meeting of opposites in life. Castorp imagines that, like the sand crystals on the beach, the snowflakes – which he catches on the sleeve of his coat, like Kepler – are made up of tiny, tiny stars, hexagons, almost identical, yet different, with infinite diversity. Deep down, the whole world, the 'source of protoplasm' as he calls it, including the human body, is made up of such crystals. A frighteningly perfect regularity, Castorp thinks. The crystals are 'the equilateral, equiangled hexagon . . . life-denying . . . icily regular in form'; 'the living principle shuddered at such perfect precision, found it deathly, the very marrow of death'. Individually, they have nothing to tell us. It is when they are brought together into a whole, forming one element, that they acquire meaning.

All in all, the encounter with the snow – 'Snow' is also the title of this thirty-five-page section in chapter six of the novel – seems to have created, or rather consolidated, a prominent sense in Castorp of what might be called dependence on the elements. Which thus manifests itself as kinship.

Wandering in the snow, the engineer discovers that precision and structure, however fascinating and even beautiful, depend on the power of life. The beating of the heart, the plasma of all living things, adds something that is a primary condition. Something that the individual crystals alone cannot produce or explain. In the face of this force, reason itself falls short. In the end, the only thing that helps is love, which is the emotion that is connected with life, its sustenance and its allurement. The insight has left Castorp rapt and warm when he finally wakes up from his dream and brushes off the snow: 'My heart beats high and knows why. It beats not solely on physical grounds . . . but humanly, on grounds of my joyful spirits.'

After his epiphany, Castorp quickly skis back to the Berghof of those marked by death, without getting an inch off-course. He is a great insight richer, and more alive than ever. In the buzzing dining room, he resumes his conversation with the Kyrgyz lady Claudia Chauchat and continues to admire her arms, draped for the evening in black silk lace with a tight row of buttons at the wrist . . .

The first decades of the twentieth century produced a line-up of snow prophets. Men like Paulcke, Bilgeri and Nansen. Only men. They had many differences, but also similarities. They look like a particular *type* representing the breakthrough era of skiing. But, more than that, a type of the new era, too, in which there's a lot of recreation taking place in the snow, including a wide-ranging array of new activities: tourism, sport, defence, risk assessment and safety efforts, science, technology, politics, nationalizing the masses using nature as a tool. Even snow. They were the type of people who took on big roles. That made them versatile. You could say they embodied a kind of synthesis. They produced combinations of the many civil projects going on within snow environments. Their projects and careers reflected how a collection of ambitions could come together within them as individuals.

They represent a new chapter in the history of snow, though one that came to a rather quick close. This shift took place at differing speeds in the different countries, but Switzerland was key. In stark terms, what happened was that snow science evolved into a *laboratory* science making universal claims that were relevant everywhere and, as such, ideal for export. A new model developed around strategic thinking on snow and security, and the implications of that model were profound.

What began the change in the way that snow and avalanche research was conducted was the fact that the model of nature as a laboratory seemed to have reached the end of the road. Paulcke himself began to doubt his model, the concept of a national outdoor laboratory with citizen scientists. As a central figure in the field of avalanche research, having fresh insight from successful experimental water research, he began to advocate for an international laboratory for snow and avalanche research.

He took the matter up with the Swiss Federal Railway in February of 1931, but met with a chilly response. The idea got more traction with forestry authorities, and contacts were made with the major technical university (ETH) in Zurich. A meeting took place in December of the same year, attended by hydroelectric companies and private railway companies. Once again, though, the conclusion was that a laboratory was not the right path to take. 'No matter what the circumstance, we should refrain from building a laboratory.' But they did decide to establish the Commission for Avalanche Research, instead. That might seem like a small thing, but it was to prove decisive. The members of this commission were themselves scientists, and they rejected the idea of recruiting both foreign experts, such as the German Paulcke, and the kind of older avalanche experts who'd relied mainly on amateur field observations, seldom visiting the Alpine valleys in winter themselves.

The Swiss experts wanted to modernize and make avalanche knowledge more scientific, so they started a form of crowdfunding. Energy companies, railway companies and the

Swiss federal postal office were invited to participate, along with anyone else who wanted to contribute. The commission also ensured that research stations throughout the Alps were in contact with each other and started to coordinate their work. The focus was on avalanche research, but as the work progressed, it became clear that more fundamental scientific questions needed to be asked – particularly concerning the properties and structure of snow. Why does the snow in a particular place sometimes become an avalanche and sometimes not? After two years of work, they changed their name to the *Expertenkommission für Schnee- und Lawinenforschung* (Expert Commission for Snow and Avalanche Study). They needed to know more about snow.

Just like Ukichiro Nakaya in Hokkaido, Japan, they took an experimental approach to the fundamentals. The physicist Henri Bader was recruited to be their crystallographer; the rest were engineers and hydrologists. And just as with Nakaya's team, none of them had even the slightest experience in researching snow. They treated snow like any other subject they might study. The assignment was to determine, without any preconceived notions, what sort of substance this really was. To investigate its properties under varying conditions.

An initial step was to find out more about the snow layering that had long been universally thought to play a major role in the formation of avalanches. Could that be confirmed? And if so, was there more to be gleaned about it? The impetus for studying snow stratification emerged out of Ernst Sorge's work at *Eismitte* on Greenland during the winter of 1930–31, the results of which became widely known the following summer, at the beginning of the second International Polar

Year, 1932–33. At the same time, a major shift was taking place across polar research. The previously dominant practice of expedition research had started showing signs that it was no longer providing the desired results. One person thinking along these lines was the Swedish geographer Hans Ahlmann. In advance of the upcoming Polar Year, he wrote a programmatic article on the need to raise expectations about what results should be returned. Ahlmann wrote that there 'is no polar research', intentionally overstating his case. What he meant was that it was the results themselves that were important, regardless of how researchers achieved them. The same methodological demands should be placed upon anyone doing observations and data collection in the field as they were in a well-equipped laboratory. The expeditions often had too many unrelated motivations and too many amateurs were involved.

This kind of criticism became more common in the interwar period, after there had been several tragic and unsuccessful expeditions while, at the same time, the experimental sciences were consistently breaking new ground. This led to a swing in the other direction. So, while the Swiss snow research was pioneering, it also reflected a more general shift in the geophysical field sciences. But the Swiss moved quickly and soon established a field research laboratory right in the heart of snow tourism.

The first research station went into operation in December of 1935. Snow was extremely difficult to transport so, rather than having to move it to some perfectly equipped laboratory in some lowland town many hours away, the station was built in Davos village, amid the snow itself, as had been the case in Hokkaido. It was also built entirely of snow and measured just 4 x 3 x 2 metres – a minimalist box, modernist and contemporary. Valuable results were already delivered in the station's first winter, and everyone agreed that the next step was to build a standard research station in a location that had been more specifically selected. The chosen location was Weissfluhjoch, at an elevation of more than a thousand metres above Davos village, but easily accessible by mountain railway. Operations at the new research station began in 1936. Now the research programme could expand, using new instruments out in the field, as well as others inside the station under controlled conditions. A new understanding of snow emerged that was virtually unrivalled anywhere in the world. While the work followed on from a decades-old tradition of studying snow and avalanches in Switzerland, it was done using completely new methods. Snow, long the subject of ad hoc investigation, was now finally made scientific.

Another similarity with Nakaya and his team at Hokkaido was the time it took from the conceptualization of a new research programme to getting results, which was about five years. The first Swiss programme lasted from 1934 to 1938, and in 1939 the flagship book *Der Schnee und seine Metamorphose* ('Snow and its Metamorphism') was published, summarizing the primary results. The key term in the book, for which Henri Bader was the primary author, was 'metamorphoses'. That's the natural condition of snow – transformation. Even on its journey down to earth, snow is transformed, then it enters into a constant state of transformation, an endless metamorphosis. A flake from the sky today is related to the snow that fell on Snowball Earth 2.5 billion years ago, and since that time it's gone through every state a water molecule can achieve.

This was the first time anyone had used the word 'metamorphosis' so extensively. Bader and his research team drew on the long domestic tradition of snow knowledge and quoted extensively from what was now a sizable body of literature about snow. But their work was also a clear step forward towards a more universal and professional understanding of snow. The beginning of a new phase. The chapter on avalanche formation surpassed anything the world had seen before. Every trace of anything metaphysical was eradicated. And there wasn't much content about life-saving and equipment, even less about *Stemmbogen* (the snowplough manoeuvre) and the appropriate length and use of ski poles. This was the opposite of Georg Bilgeri's approach. The net effects of the avalanches received little attention, although the book shows some pictures of the Johann Coaz-designed avalanche barriers. Instead, *Der Schnee und seine Metamorphose* was a report about snow. The main focus was on investigating its behaviour. In order to create protection from avalanches, to predict them, to assess the risks, to build barriers against them – because that was the main aim – then what is needed, first and foremost, is to understand the mechanics, dynamics and plasticity of snow.

An avalanche itself is also, in a way, a metamorphosis. It was implicit in the title's *'Metamorphose'* – that is, *plural*. The book explained that it began in the metamorphosis of the snow crystal itself – the many snowflakes that sublimated and vanished before reaching the ground – continuing through countless crystalline transformations, including stratification, firn, glaciation and annihilation under heat.

No one had really talked about snow like that before. Perhaps it was only here that the snowflake, in the long tradition from Kepler and Descartes to Nakaya, was finally allowed to join the grand scale of snow in the landscape. Flake with snowfield, flake with ice sheet, flake with rising sea levels.

Davos and Hokkaido, Bader and Nakaya – the world's leading snow experts outside of the Soviet Union now joined hands. Crystallographer Henri Bader, hydrologist Robert Haefeli, geologist Paul Niggli and their colleagues quoted, with approval, Nakaya and his colleagues when they wrote about 'The Physics of Skiing'. Bader adopted Nakaya's snow classification, word for word; he just translated it into German. Nakaya, in turn, quoted from Bader. The entire 1939 report was translated into English, but not until 1954, by which time the Americans had begun to realize the importance of knowledge about snow and how much it was needed.

In the late 1940s, they approached André Roch, another Swiss national with experience from the snow laboratory in Davos, where he had been head of avalanche research. Moreover, he had experience of working in the United States, as a ski instructor, avalanche expert and developer of winter sports resorts, including Aspen. Bader himself had come to the United States immediately after the war, becoming in 1954 director of the Snow, Ice and Permafrost Research Establishment (SIPRE). During the war years, he'd first taught in Bogotá, the capital of Colombia, before becoming director of a mine in Curaçao. In the summer of 1945, he became a professor at Rutgers University in New Jersey. After a few years, he was commissioned by the US government to investigate the state of glaciology in key European countries, and compare it with that in the United States. The investigation revealed serious shortcomings on the American side. After the war, as we have seen, the Nakaya and Bader teams would converge at the Bader Institute in the USA, and on the Greenland Ice Sheet.

On this point there was clear resonance between the Sapporo and Davos laboratories. They followed totally different research traditions – Nakaya with his snowflakes in the lab, Bader and his colleagues with their measuring instruments out in the snowy terrain. But they increasingly realized that, together, they had access to knowledge that was indispensable for anyone wanting to understand snow and reduce the danger from it. What set their knowledge apart from much of the past was that they understood snow as a substance, from the inside out. That's what the Americans needed, and why they had recruited Henri Bader to be their chief scientist. The Americans also had ambitious publication and translation programmes, and published bibliographic summaries of current snow research. Bader, in turn, reached out to Japan and Ukichiro Nakaya. That's also why, after years of working mostly on their own in the Swiss Alps and the mountains of Hokkaido respectively, the two traditions of snow crystal research converged in American laboratories and on the Greenland ice, propelled by new geopolitical conditions in the Cold War.

One publication that demonstrated the growing collaboration was *The Physics and Mechanics of Snow as a Material* from 1962, written largely by Bader, alongside a Japanese snow scientist, Daisuke Kuroiwa, who also contributed a section on the electrical properties of snow. The report was part of a larger research and publication programme conducted under the auspices of the US Army Cold Regions Research and Engineering Laboratory (CRREL) in Hanover, New Hampshire – nothing was left to chance. Knowledge was put to use, wherever it could be found. This follows a major pattern that the US applied throughout the Cold War – that of using its strong position and key role in security policy, not least within NATO, to enlist the capabilities and expertise of other countries.

One such area concerned the environment as a critical factor in human performance, especially in warfare. This wasn't unusual. It had been well known since time immemorial that people at high altitudes had a harder time breathing and could get less done. Humid, hot environments were also challenging for taxing work. In cold environments, which are often also dark and snowy, it's difficult to function effectively. People who aren't acclimatized to it can suffer from depression and anxiety. If you're sweating for a long time, you'll eventually become tired and confused. If you don't sleep well, you function less well across the board. The research – biological, physiological, psychological and anthropological – that deepened our knowledge of these phenomena was of great interest not only to the military, but also to industry.

Two simple ideas emerged. One was that if you can understand how people are affected by their local environment, you can prepare them, making them better able to cope with stress, or redesign the environment itself to make it less strenuous. When high-altitude jet aircraft were first developed during the Second World War for military use, there was an urgent need to learn more about pressure changes and to design cockpits with an appropriate 'environment'. There was also a need to develop methods for training pilots to do precision work under atmospheric pressure, or in confined spaces that could easily trigger feelings of stress and panic. The second idea was that they needed materials and methods that reduced the risks of environmental impact. Was it possible to research properties for socks and shoes that reduced the risk of chafing? Could textiles be developed that didn't absorb as much body fluid or that 'breathed' to reduce perspiration? Could food rations be developed that were perfectly balanced for use in the field and processed to optimize the amount of energy needed relative to weight? What exactly was an optimal breakfast for physical performance?

There were no well-thought-out answers to these kinds of questions. In the United States, several federal, publicly funded laboratories were set up for such 'environmental research', including one focusing on snow and ice at the University of Fairbanks in Alaska – the Arctic was a potential theatre for a third world war. Meanwhile, a special 'fatigue laboratory' was established at Harvard University in Boston. The research there was conducted between 1927 and the facility's closure in 1947, and included both high-altitude research – with expeditions to high summits in the Andes – and stress tests in both cold and tropical environments. There was also extensive materials development.

Some of the Harvard Fatigue Laboratory's most dramatic experiments were conducted in Arctic environments, including experiments with low-calorie diets to see what happened to soldiers' performance under starvation: it nosedived. Great interest was shown in the materials and techniques of the Arctic peoples. At the time, the properties of their clothing and footwear made from fur and leather couldn't be improved upon using artificial materials.

Similar research was carried out in several countries. In Sweden, the most advanced performance-physiology research was conducted at the Royal Gymnastics Central Institute (GCI) in Stockholm (now the Swedish School of Sport and Health Sciences). There were applications for snow here, too, for example when Per-Olof Åstrand and his future wife Irma Ryhming introduced a static bicycle measuring oxygen uptake, for the Swedish national ski teams in the early 1950s. You could train scientifically, was their message. The skiers, who'd been trained in forestry work under snow-covered spruces, were sceptical. Someone asked Åstrand if he had his wires crossed – this was the national ski team, not the national cycling team. But soon enough, 'rational training' was everyone's catchphrase. The test cycle was introduced by the first professor of performance physiology at the GCI, Erik Hohwü-Christensen from Denmark. He gained his expertise from his teacher August Krogh (also Danish), winner of the 1920 Nobel Prize in Medicine. Krogh had studied respiration and blood flow, and had been able to clarify in detail how oxygen is absorbed in the body. His early research had centred on the exchange of carbon dioxide between the air and the sea – which he had studied in Greenland.

It transpires that there's an oxygen–carbon dioxide exchange taking place at the individual human level, and another on a planetary scale. Even beneath and inside the snow, the earth's never-ending breathing was taking place. Everything rested on a delicate balance, in which the availability of oxygen was just as crucial as the limit to the amount of carbon dioxide. If these exchanges were disrupted, if respiration didn't work properly, if the oceans couldn't absorb the carbon dioxide, then heat would rise in the atmosphere. Irish physicist John Tyndall and the Swedish chemist Svante Arrhenius had already realized this back in the nineteenth century. The consequences of an imbalance – for the planet or the human body – would be equally devastating. On an overheated Earth, the systems that were needed to support life would not work. And if the oxygen supply was too weak in the cockpit, the pilot would lose the battle.

Research on snow and the environment was generating interconnected insights that would prove vital in the fossil-fuelled world taking shape after the Second World War. But these insights hadn't yet informed each other closely enough. The research was being conducted in different environments, in different places, with differing methodologies, all with the new Iron Curtain erecting an unforeseen Cold War barrier to the exchange of knowledge. This environmental research on a micro level, on equipment and physical performance, had a counterpart for the macro-environment, if we can call it that. Military interests were even more dominant in that space in the beginning. During the Second World War, it had become clear that modern major warfare could take place over a vast geographical area and required rapid communication across virtually the entire globe. It required infrastructure and equipment. It was also fought in very different places, or 'environments': on land, at sea and in the air. The war also covered a wide range of climates: tropical jungles, tundra and glaciers, hot deserts, rugged mountain landscapes, swamps and forests.

But war also had a vertical dimension. On the drawing board at the Pentagon and in the military-funded research institutes there were new vehicles (satellites, submarines for operations under the Arctic ice, entirely new types of air-craft) to master the new environments. But research didn't stop there. It increasingly centred on the characteristics of the macro-environments that could constitute the theatres of war of the future. These went right up into the strato-sphere, they were deep below the surface of the oceans and they penetrated the subsoil. They were on the vast snowfields of glaciers. They were also indoors, in bunkers and battle-management centres. And they were inside people, in their feelings and thoughts, often obscure, sometimes dark.

The word used both for the human environment (even at the level of underwear, socks) and for the very large planetary regions (rainforest, sea ice, tundra, desert, steppe) eventu-ally became the 'environment', which began its phenomenal career as a concept in the early postwar years. Nowadays, ice and snow are equally obvious parts of that environment. They are indicators of environmental change, and we recog-nize that snow and ice are therefore a global element – but they weren't thought of that way when Hans Ahlmann vis-ited the Pentagon in May of 1947 to share his findings. Snow was more than words, more than a substance, more than crystals. The cold was strategic, and snow had been given a role in the ongoing transformation of the world, for better or worse.

How this all fitted together was still not completely understood. But more and more people began to recog-nize that it was important, and that it required the kind of understanding that could see the planet's various processes as part of a composite whole. A powerful, finely calibrated apparatus – or maybe, better yet, an *organism*. Cold and heat, they realized, were key categories. How cold or warm the planet was made all the difference for life on Earth. And that included human life. If we were going to be able to breathe, the whole planet had to be able to breathe. And on a breathable planet, there's snow. If the snow stops coming and going with the seasons, like an annual inhalation and exhalation of whiteness and restfulness over the earth – then our own, much faster breathing isn't going to function either. It was this idea that began to take shape, somehow travelling wordlessly out into the world like oxygen molecules themselves. Something that also animated the uniquely progressive and science-affirming snow-romance of the post-war period. My childhood terry-cloth shower towel with its picture of an igloo held a meaning that I am only now able to understand.

Afterword

The Angels' Conversation

SNOW IN SWEDEN HAS ALWAYS BEEN PRAISED for its usefulness. It's significant that Carl Linnaeus's reflections on snow, now nearly three hundred years ago, were about its economic advantages. That makes him typically Swedish. Swedes are factual; they like having a non-mystical world. Even the very snow itself has to be useful.

My grandfather, a politician, is an example of this. He had the best of intentions. He'd experienced poverty, and he knew how important it was to break that cycle. But he didn't know much about what might be possible with prosperity. He'd had no formal education and never read a book. There was quite a lot that he couldn't imagine.

Those of us living now can reach further than that. Along paths that we forge for ourselves, each and every one of us. It's about following your own inner compass.

'God send everyone their heart's desire', as Shakespeare wrote.

It's November 1981 and I'm sitting on a train pushing patiently forward through the already considerable snowbanks of the innermost region of Norrland. I'm on my way home after a long trip. I've stood on the border of apartheid

South Africa, where I was by no means welcome. I've met the kind of poverty and complete absence of education that the Portuguese had left behind in Mozambique.

The world I met in Africa had been filled with sound – sounds of human protest, of the shrieks of nocturnal animals, the howling alarm of history, the raging storm blowing out of Paradise towards the Angel of History with the wide-open eyes that can never shut. It is the exact opposite when the train I am travelling on comes to a stop along the little track through Ångermanland and Västerbotten. It is completely silent.

The silence affects me. It makes me feel small. Returns the world to its rightful scale. And makes me humble, a word I learned as a child but is now seldom heard. It was as if the snow was saying, *We can change. Start over.*

That thought has never left me.

Now snow is a part of the Anthropocene. One of the most fleeting substances in existence. Hovering around the zero-degree line. Yet it's everywhere. The ultimate test for the Paris Agreement to reduce global greenhouse-gas emissions to safe levels. An indicator of the health of our civilization, the capacity of our societies, our resolve as individuals.

It is morality, written in diamonds.

This is very difficult. We're collectively rounded up into societies which to this point have shown themselves incapable of doing what we ought to do. What we committed to doing, what we promised ourselves and each other. We are caught up in circumstances that we don't have the capacity to get out of: fear, routine, explaining why we aren't able to. Those who ought to possess the resolve to do what needs to be done, completely lack that resolve. What they say they don't want to do is exactly what they're doing.

Anyone whose actions leave the snow melting, and the glaciers dying, will likely never be held accountable. Those who want to live righteously are seldom rewarded.

But it's right to try.

Snow angels. Can these be made with words? If so, I'd like to think of this book as an attempt.

I wrote at the outset that I wanted to awaken the snow angel within us. What did I mean? I believe that within each and every one of us there's a better self that sees what's happening and whose emotions awaken as it becomes apparent that the snow on the ground beneath our feet is disappearing. That's our love of the snow making itself known. That's good, because that makes more of us who care about the snow. Personally, I want the snow to stay the way we remember it, continuing to light up the dark.

The snow angel reminds me of the candles I sometimes buy and light when I enter a medieval cathedral somewhere in Europe. I am making a small contribution to the survival of the sanctuary. But I am also making a gesture in favour of something that has existed for a long time, something that I hope will continue.

It's possible to conceive of a world without snow, of course. It's probably possible to live in it, too. But for me and certainly for many others, it would be a more dangerous world – and a poorer life. We can think of that relationship as a belief. A faith that is not a presumption of truth. Not 'what I believe is true' – but 'what I believe is what I deeply and sincerely value'.

My belief is that snow means something to us, even collectively. Snow is part of the breathing of the world. It keeps us alive. It also keeps hope alive.

Snow is an entity that comes into our lives from time to time, without prior warning, telling us in its wordless way that the world can be different. That it can be transformed.

The snow is a testimony to what I call grace. Not an assurance of salvation, but a summons to responsibility.

The snow angel represents the opportunity for moral action. If it has a face, I believe it is directed towards the future. It's an angel that can help us see what we ought to be doing.

The Angel of History, on the other hand, is condemned to look backwards upon the growing ruins of the past. It is now over a century since Walter Benjamin acquired Paul Klee's picture, and nearly as long since he wrote the philosophical-historical 'Thesis IX' of his *Theses on the Philosophy of History* (1939), in which he explained what he meant by the Angel of History. Benjamin emphasized that it was more important to remember the damage people had done to each other in the past than to call for reconciliation and progress in the future.

The snow angel isn't an attempt to soften Walter Benjamin's concept. Without the memory of injustice, and having empathy with that suffering, the celebration of progress becomes a false idyll. The singular contribution of the snow angel is that it reminds us of something that didn't seem to belong to history when Benjamin was alive, and which he therefore could not imagine as part of the past that the angel would witness during his backward, storm-cast flight into the future: the snow itself, the elemental substance and what snow is a part of – the elements of air, earth, water, ice, forests, wetlands and the material fabric of life. Things that were once thought to stand outside human history, but which, within the past century, have become part of it.

I want to appoint the snow angel as a guardian angel for these destroyed and deformed elements. First and foremost to acknowledge their rightful place among the other horrific occurrences in history's enormous litany of destruction.

Snow is an apt form for this task. Not only because of its angelic colour, but also because of the important role it plays in the elemental cycle. It connects everything on its quest from the sky down to the earth, and in its ongoing travel through glaciers and snowfields, rivers and lakes, all the way out to irrigating the soil and ecosystems, villages and communities. Snow is a waystation for water and a brake on the pace of the earth. It's also the earth's mirror, whose whiteness reflects the sun's energy back out into space, cooling our hot planet.

That's why snow is so well suited to the role of protector. It's a forgiving element that helps flawed people. By taking good care of snow, we can mitigate the devastation we're causing.

In my mind's eye, I see how the two angels begin a conversation. The Angel of Snow is telling the Angel of History about the place elements have taken in the past. She explains why in recent times the elements have become such a large new part of the 'pile of ruins', which it falls to the Angel of History to contemplate without respite, as the pile 'grows up to heaven'. She explains why these ruins of nature – because the elements were once nature – join the ranks of murdered and tortured people, the poor and displaced, and the unjustly imprisoned. And why, superimposed over the deathscapes of war and plague, there are the extinction of species, the destruction of rainforests, the collapse of the tundra and an atmosphere steeped in carbon dioxide.

Ever more frequently, the Angel of History has to yell backwards at his colleague, the Angel of Snow looking forward: 'I see now that you're completely right!'

And for as long as this continues, he has to integrate the elements, even snow, into part of his image of the history that humanity creates. So these angels travel through time, back-to-back like conjoined twins of our late Anthropocene modernity, in a moral alliance, with the mission of helping us humans, with our limited vision and poor abilities. Helping us to remember and understand, and to let hope and forgiveness guide our actions here on earth. Every time it is said, it sounds like a curse. Yet I don't understand what else there is to be said.

In his 'Thesis VI', Walter Benjamin writes, 'To articulate the past historically does not mean to recognize it "as it really was".' Just the opposite, Benjamin writes: it means 'grasping memory as it arises in a moment of danger'. I want us to remember the snow and feel it, so that we realize that we're in just such a moment of danger.

In this way, seeing the dangers of the present is a way to discover a new history. As a historical category, snow does not exist until it is realized by the present. Progress isn't a road forward to a new paradise that we're collectively hoping to reach. That's a misunderstanding. To see progress as the achievement of a predetermined goal is to trivialize history and de-historicize the present.

The history of snow is instead part of a history *in the*

making, a history we've never possessed before. It is my belief that this will grow and become a greater history of the elements, which will contribute to the redefining of humanity's image of itself on earth and, accordingly, our image of what true *progress* can be. This beautiful word of obligation, which in our era needs a reimagining.

The snow we have, and even the snow we no longer have, depends more and more on us ourselves. Snow's presence in the future is our responsibility now. The snow angel can become our symbol for how we take this responsibility.

I wanted the two angels – of History and of Snow – to start talking to each other. I guess that's why I always wanted to write this book.

Selected References

All website links active as of April 2025.

INTRODUCTION: SNOW ANGEL

Angel of History: Walter Benjamin, 'Theses on the philosophy of history', in *Illuminations*, ed. H. Arendt, tr. H. Zohn (New York: Schocken Books, 1969).

Constellations, moral responsibility: Immanuel Kant (1781), *Kritik der reinen Vernunft* [Critique of pure reason].

Sisyphus's happiness: Albert Camus (1942), *The Myth of Sisyphus* (London: Penguin Classics, 2013).

Snow as a symbol: Thomas Seiler, *Snøens formler: Et naturfenomen i litteraturen* [The formulas of snow: a natural phenomenon in literature] (Oslo: Cappelen Damm, 2023).

Earth's first snowfall: I. N. Bindeman et al., 'Rapid emergence of subaerial landmasses and onset of a modern hydrologic cycle 2.5 billion years ago', *Nature*, 557:7706 (2018), pp. 545–8.

Skilled skiers: D. Svensson, 'Scientizing performance in endurance sports: The emergence of "rational training" in cross-country skiing, 1930–1980', PhD thesis (Stockholm: KTH, 2016).

Swedish snow records: L. Wern, 'Snödjup i Sverige [Snow depth in Sweden] 1904/05–2013/14', report 158 (Swedish Meteorological and Hydrological Institute, 2015).

Disappearance of snow: M. Carrer et al., 'Recent waning snowpack in the Alps is unprecedented in the last six centuries', *Nature Climate Change*, 13 (2023), 155–160; A. R. Contosta et al., 'Northern forest winters have lost cold, snowy conditions that are important for ecosystems and human communities', *Ecological Applications*, 29:7 (2019), pp. e01974.

Answers about snow: Yngve Ryd, *Snö: Renskötaren Johan Rassa berättar* [Snow: In the words of reindeer herder Johan Rassa] (1999) (Stockholm: Natur & Kultur, 2001), p. 7.

Monkeys make snowballs: Ruth Kirk (1977), *Snow* (Seattle and London: University of Washington, 1998), p. 7.

DIAMONDS AND BIRDS' WINGS

History of the word 'snow': 'Snow', Wiktionary, https://en.wiktionary.org/wiki/snow; 'Snö', *Swedish Academy Dictionary* (SOAB), https://www.saob.se/artikel/?unik=S_08241-0204.Wn7N&pz=3; Elof Hellquist (1922), *Svensk etymologisk ordbok* [Swedish etymological dictionary] (Lund: LiberLäromedel/Gleerup, 1980), vol. 2, p. 1021.

Snow's metamorphoses: Henri Bader et al., *Der Schnee und seine Metamorphose* [Snow and its metamorphoses] (Bern: Kümmerly and Frey, 1939).

Lapphandskar: F. V. Svenonius, 'Det snöar lapphandskar' [It's snowing Lapp gloves], photograph, *Alvin*, 20 Feb. 1889, https://www.alvin-portal.org/alvin/view.jsf?pid=alvin-record%3A269854&dswid=987

Emergence of the expression: Petrus Læstadius, *Fortsättning af journalen öfver missions-resor i Lappmarken Innefattande åren 1828–1832* [Continuation of the journal of missionary travels in Lapland during the years 1828–1832] (Stockholm: Henr. Gust. Nordström, 1833), p. 452.

World's largest snowflake: Guinness World Records, 'Largest snowflake', https://www.guinnessworldrecords.com/world-records/73325-largest-snowflake

Big snowflakes in Umeå: J. Lindberg, 'Gigantiska snöflingor över Umeå på torsdagen' [Giant snowflakes over Umeå on Thursday], *Västerbottens-Kuriren*, 14 March 2024.

World's largest hailstone: National Climate Extremes Committee, 'Memorandum for the record', 1 Nov. 2010, https://www.ncei.noaa.gov/monitoring-content/extremes/ncec/reports/vivian-hailstone-final.pdf

Much larger hail: Évariste Régis Huc, *Travels in Tartary, Thibet, and China During the Years 1844-5-6*, tr. W. Hazlitt (London: Vizetelly and Co., 1851), vol. 1, p. 12.

Missing inland ice: 'McMurdo Dry Valleys', *Wikipedia*, https://en.wikipedia.org/wiki/McMurdo_Dry_Valleys

Snowfall in Skåne: L. Wern, 'Snödjup i Sverige [Snow depth in Sweden] 1904/05–2013/14', report 158 (Swedish Meteorological and Hydrological Institute, 2015). For further data on extreme snowfall and snow depth worldwide, see C. Dolce, '5 incredible snowfall extremes', Weather Channel, 7 Jan. 2015, https://weather.com/safety/winter/news/five-snowfall-extremes-20130103; Salam Groovy Japan, 'Japan is the world's snowiest country, 51% covered by snow', https://www.groovyjapan.com/en/japan-snow/. There is also interesting if unverified data at C. C. Burt, 'U. S. Snowfall 1900–2019: A decade-by-decade look', *Weather Underground*, 3 Jan. 2020, https://www.wunderground.com/cat6/US-Snowfall-1900-2019-Decade-Decade-Look

Types of snow in Sámi: Yngve Ryd (1999), *Snö: Renskötaren Johan Rassa berättar* [Snow: In the words of reindeer herder Johan Rassa] (Stockholm: Natur & Kultur, 2001), pp. 43, 44, 83.

Reindeer as forest livestock: Kjell-Åke Aronsson, *Forest Reindeer Herding A.D. 1–1800: An Archaeological and Palaeoecological Study in Northern Sweden* (Umeå University Department of Archaeology, 1991).

Terms *skare* and *klabbföre*: see respective definitions in *Swedish Academy Dictionary* (SAOB), https://www.saob.se

Buffon: Georges-Louis Leclerc (1778), *The Epochs of Nature*, tr. and ed. J. Zalasiewicz, A.-S. Milon and M. Zalasiewicz (University of Chicago Press, 2018).

SNOW ANXIETY

Glaciers are dying: M. Färnbo, 'Gravsten över "död glaciär" har ett ödesmättat budskap om framtiden' [Epitaph over 'dead glacier' holds ominous message about the future], *OmVärlden*, 19 Aug. 2019.

Iceland's glaciers are going to melt: 'The entrance to the "centre of the earth" may disappear within decades', United Nations Regional Information Centre for Western Europe, 10 Feb. 2023, https://unric.org/en/the-glaciers-in-iceland-may-disappear-within-decades/

Emergence of the 'Holocene' concept: P. Gervais, 'Sur la répartition des mammifères fossiles entre les différents étages tertiaires qui concourent à former le sol de la France' [On the distribution of fossil mammals between the different Tertiary stages which contribute to forming the soil of France], *Académie des Sciences et Lettres de Montpellier*, 28:21 (1850), pp. 399–413.

Ice archive: M. Walker et al., 'Formal ratification of the subdivision of the Holocene Series/Epoch (Quaternary System/Period): two new Global Boundary Stratotype Sections and Points (GSSPs) and three new stages/subseries', *Episodes*, 41:4 (2018), pp. 213–23; E. Kolbert, 'The "epic row" over a new epoch', *New Yorker*, 20 April 2024. The ice core archive can be seen at https://www.isogklima.nbi.ku.dk/nyhedsfolder/uk_with_dk_companion/modern_age_is_defined_kopi/

Human timescale: Sverker Sörlin (2017), *Antropocen: En essä om människans*

tidsålder [The Anthropocene: An essay on the age of humans] (Stockholm: Weyler, 2024).

Just the beginning: R. Lindsey and L. Dahlmann, 'Climate change: global temperature', Climate.gov, 18 Jan. 2018, https://www.climate.gov/news-features/understanding-climate/climate-change-global-temperature

Goodbye to Pizol: 'Schweizisk sorgemarsch för smältande glaciär' [Swiss funeral march for a melting glacier], *Syre*, 23 Sept. 2019.

Oregon glacier: E. Bouhassira, 'Clark glacier commemorated at funeral in Oregon', *State of the Planet* (*Columbia Climate School*), 27 Oct. 2020, https://news.climate.columbia.edu/2020/10/27/clark-glacier-funeral-oregon/

Loss of the Humboldt Glacier: N. Vallangi, 'Venezuela loses its last glacier as it shrinks down to an ice field', *Guardian*, 8 May 2024.

Snow disappearing is a new phenomenon: K. E. Kunkel et al., 'Trends and extremes in northern hemisphere snow characteristics', *Current Climate Change Reports*, 2 (2016), pp. 65–73. The period between 1967 and 2023 shows a consistent decline in the northern hemisphere winter snow cover of 10 per cent and a decline in the annual number of days with snow cover by an average of three days per decade between 1972 and 2022 – see 'Snow: why it matters', National Snow and Ice Data Center, https://nsidc.org/learn/parts-cryosphere/snow/why-snow-matters#anchor-impact-of-climate-change-on-snow

Nordic folk and Britons would get a little sun: S. Sörlin, 'The global warming that did not happen: Historicizing glaciology and climate change', in S. Sörlin and P. Warde (eds), *Nature's End: History and the Environment* (London: Palgrave, 2009), pp. 93–114.

Hans Ahlmann: 'Den aktuella klimatförbättringen' [The current climate improvement], *Svenska Dagbladet*, 16 Sept. 1938.

Enthusiasm for climate improvement: 'Det var kallare förr – läs här om klimatförbättringen: Fabrikerna ger oss mildare klimat: Det är kolsyran som gör'et' [It was colder in the past – read this about climate improvement: factories give us a warmer climate: CO_2 does the trick], *Aftonbladet*, 14 March 1954.

AMOR NIVIS

Wilcke's ice hexagons: J. C. Wilcke, 'Rön och tankar om snö-figurers skiljaktighet' [Thoughts and opinions on the disparity in snow forms], *Kongl. Vetenskapsakademiens Handlingar* [Royal Swedish Academy of Sciences Proceedings] (*KVAH*), 22 (1761), pp. 1–19; Wilcke, 'Nya rön om vattnets frysning til snölike is-figurer' [New thinking about how water freezes into snow-like ice shapes], *KVAH*, 30 (1769), pp. 90–111; Wilcke, 'Om snöns kyla vid smältningen' [About how cold snow is as it melts], *KVAH*, 33 (1772), pp. 97–120.

Reasons for the crisis from the past: R. Koselleck, 'Crisis', *Journal of the History of Ideas*, 67:2 (2006), pp. 357–400; Koselleck, '"Erfahrungsraum" und "Erwartungshorizont": zwei historische Kategorien', *Vergangene Zukunft* ["Experiential space" and "expectation horizon": two historical categories, past future] (Frankfurt am Main: Suhrkamp, 1979).

'The most astonishing event': Edmund Burke, *Reflections on the Revolution in France, and on the Proceedings in Certain Societies in London Relative to That Event* (London: J. Dodsley, 1790) p. 11.

Dante's third circle: Dante Alighieri, *The Divine Comedy*, tr. Allen Mandelbaum (London: Everyman's Library, 1995), *Inferno*, canto VI, lines 10–12.

Swedish literary scholar: Martin Lamm, *Upplysningstidens romantik: Den mystiskt sentimentala strömningen i svensk litteratur* [The romanticism of the Enlightenment: The mystic-sentimental current in Swedish literature], 2 vols. (Stockholm: Geber, 1918–1920).

Middle-class family culture: Eric Johannesson, *Den läsande familjen: Familjetidskriften i Sverige 1850–1880.* [The reading family: The family magazine in Sweden 1850–1880] (Stockholm: Nordiska Museet, 1980).

Felix Körling: 'Felix Körling (1864–1937)', Swedish Musical Heritage, https://levandemusikarv.se/composers/korling-felix/

Snow insulates: Martin Jansson, 'Om värmeledningsförmågan hos snö' [On the thermal conductivity of snow] (Uppsala: Akademiska Boktryckeriet, 1904).

Research on snow: T. V. Callaghan et al., 'The changing face of Arctic snow cover: A synthesis of observed and projected changes', *Ambio*, 40 (2011), suppl. 1, pp. 17–31.

Cultural history of the Vasaloppet ski race: D. Svensson, 'I fäders spår? Längdskidåkningens landskap som kulturarv', [In the tracks of the fathers?: The landscape of cross-country skiing as cultural heritage], *RIG*, 96:4 (2014), pp. 193–212.

Fridtjof Nansen: 'Fridtjof Nansen', *Wikipedia*, https://no.wikipedia.org/wiki/Fridtjof_Nansen; Harald Dag Jølle, *Nansen: Oppdageren* [Nansen: The Discoverer] (Oslo: Gyldendal, 2010) and *Nansen: Utfordreren* [Nansen: The Challenger] (Oslo: Gyldendal, 2020); Audun Renolen Aasbø, *Skiløperkunstens litterære kraftsentrum: Fridtjof Nansen og den moderne skiidrettens framvekst* [Literary powerhouse of the art of skiing: Fridtjof Nansen and the emergence of modern cross-country ski sport] (University of Oslo, 2014).

Lifts her out of the snow: Kari J. Spjeldnaes, *På ski fordi: Om gleden ved å gå på* *ski: Forsök till en beskrivelse* [On skis because: the joy of skiing – an attempt at a description] (Oslo: Pax, 2018).

Eva Nansen's article in defence of women skiing: Bodil Stenseth, *En norsk elite: Nasjonsbyggerne på Lysaker 1890–1940* [A Norwegian elite: The nation-builders at Lysaker: 1890–1940] (Oslo: Aschehoug, 1993), pp. 117–18.

The ideal of *friluftsliv*: Øystein Sørensen and Bo Stråth (eds), *The Cultural Construction of Norden* (Oslo: Scandinavian University Press, 1997).

'nuclear winter': P. Crutzen and J. W. Birks, 'The atmosphere after a nuclear war: twilight at noon', *Ambio*, 11 (1982), pp. 114–25; M. R. Francis, 'When Carl Sagan warned the world about nuclear winter', Smithsonianmag.com, 15 Nov. 2017, https://www.smithsonianmag.com/science-nature/when-carl-sagan-warned-world-about-nuclear-winter-180967198/; S. Turchetti, 'Trading global catastrophes: NATO's science diplomacy and nuclear winter', *Journal of Contemporary History*, 56:3 (2021), pp. 543–62.

Crutzen's research on the ozone hole: Sverker Sörlin and Eric Paglia, *Stockholm and the Rise of Global Environmental Governance* (Cambridge University Press, 2025), pp. 55–7, 83–4.

***Mystic North* exhibition:** Roald Nasgaard, *The Mystic North: Symbolist Landscape Painting in Northern Europe and North America 1890–1940* (University of Toronto Press, 1984). The Oslo National Museum provides further details in its exhibition (June 2024).

Sohlberg's paintings from Rondane: Øyvind Storm Bjerke, 'Vinternatt i fjellene' [Winter night in the mountains], in Harald Sohlberg: Infinite landscapes, National Museum, Oslo, 2018.

Ravna and Balto: 'Samuel Balto', *Norsk Polarhistorie*, https://polarhistorie. no/personer/samuel-balto/; 'Ole Nilsen Ravna', *Norsk Polarhistorie*, https:// polarhistorie.no/personer/ole-nilsen-ravna/

Snow areas equipped with skis: Tor Bomann-Larsen, *Den evige sne: En ski-historie om Norge* [The eternal snow: A skiing history of Norway] (Oslo: Cappelen, 2005), pp. 62–5.

'Skiing was completely unheard of': *Viktor Balcks minnen, 1: Ungdomen* [Viktor Balck's memories, vol. 1: Youth] (Stockholm: Bonnier, 1929).

Memoirs reinforce the image of an outdoor life: Ann Katrin Pihl Atmer, *Livet som leves där måste smaka vildmark: Sportstugor och friluftsliv 1900–1945* [The life they lived there must taste of wilderness: Sports cabins and the outdoors 1900–1945] (Stockholm: Stockholmia Förlag, 1998), pp. 45–6.

Skiing being unusual: C. J. E. Hasselberg, 'Från turistväsendets barndom: Jämtländska minnen och reseanteckningar' [From the early years of tourism:

Jämtland memories and travel notes], in *Svenska Turistföreningens Årsskrift* [Annual Report of the Swedish Tourism Association], 1934.

Norwegian stations for snow measurement: Meteorologisk Institutt, 'Station information', https://seklima.met.no/stations/

Gustaf Fjaestad: S. Sandström, 'Gustaf A. C. Fjaestad', *Svenskt Biografiskt Lexikon*, SBL 14169; Torsten Gunnarsson, *Nordic Landscape Painting in the Nineteenth Century*, tr. Nancy Adler (New Haven, Connecticut and London: Yale University Press, 1998), pp. 247–60.

Helmer Osslund: Gunnarsson, *Nordic Landscape Painting*, op. cit.

Anna Boberg: E.-C. Mebius, 'En romantisk modernist: Anna Bobergs visionära arktiska landskap' [A romantic modernist: The visionary Arctic landscapes of Anna Boberg], in *Kvinnliga pionjärer: visionära landskap* [Women pioneers: Visionary landscapes], ed. Karin Sidén (Stockholm: Prins Eugens Waldemarsudde, 2023), p. 35; Yvonne Gröning, *Fru Bob: En personlig biografi* [Mrs Bob: A personal biography] (Hedemora: Gidlunds, 2009).

Pekka Halonen: Gunnarsson, *Nordic Landscape Painting*, pp. 260–9.

Japanese influences: Annika Waenerberg, *Biografiskt lexikon för Finland, vol. 2: Ryska tiden* (Stockholm: Atlantis, 2009).

Akseli Gallen-Kallela: Juha Ilvas (ed.), *Akseli Gallen-Kallela* (Helsingfors: Ateneum, 1996).

Monet's Kolsåstoppen paintings: Nasgaard, *The Mystic North*.

Poems by Tomas Tranströmer: Tomas Tranströmer, 'C Major' and 'Snow Is Falling', in *The Great Enigma: New Collected Poems*, tr. Robin Fulton (New Directions, 2006).

CARE OF SNOW

Snow droughts can cause concern: L. S. Huning and A. AghaKouchak, 'Global snow drought hot spots and characteristics', *PNAS*, 117:33 (2020), pp. 19753–9; F. Avanzi et al., 'Winter snow deficit was a harbinger of summer 2022 socio-hydrologic drought in the Po Basin, Italy', *Communications Earth & Environment*, 5:1 (2024), p. 64.

Reduction in snow days and snow cover: L. Mudryk et al., 'Historical Northern Hemisphere snow cover trends and projected changes in the CMIP 6 multi-model ensemble', *Cryosphere*, 14 (2020), pp. 2495–514; US Environmental Protection Agency analysis of data from 2021 (9 July 2024), https://www.epa.gov/climate-indicators/climate-change-indicators-snow-cover; S. S. Young, 'Global and regional snow cover decline: 2000–2022', *Climate*, 11:8 (2023), p. 162.

Swedish winter getting shorter: J. Hövenmark, 'Klimatforskaren: Snösäsongerna blir kortare' [Climate researcher: The snow season is getting shorter], *Dagens Nyheter*, 24 Dec. 2023.

Snow in the Highlands: Iain Cameron, *The Vanishing Ice: Diaries of a Scottish Snow Hunter* (Sheffield: Vertebrate Publishing, 2021).

Feelings of loss and grief: Cameron, *The Vanishing Ice*, p. 104.

Snow stores water: P. Torralbo et al., 'Characterizing snow dynamics in semi-arid mountain regions with multitemporal Sentinel-1 imagery: A case study in the Sierra Nevada, Spain', *Remote Sensing*, 15:22 (2023), p. 5365.

Colonization by the Moors: K. Walker, 'The Moorish invention that tamed Spain's mountains', BBC, 12 Oct. 2022, https://www.bbc.co.uk/future/article/20221011-the-moorish-invention-that-tamed-spains-mountains; Fascist control of water – see Erik Swyngedouw, *Liquid Power: Contested Hydro-Modernities in Twentieth-Century Spain* (Cambridge, MA: MIT Press, 2015).

City of Granada: Manuel Titos Martínez, *Los Neveros de Sierra Nevada* [The snow workers of the Sierra Nevada] (Madrid: Organismo Autónomo de Parques Nacionales, 2014).

LIGHT AND SILENCE

***Sastrugi* snow:** Per Jansson and Per Holmlund, *Glaciologi* [Glaciology], Vetenskapsrådet & Stockholms Universitet [Swedish Research Council and Stockholm University], 2003; D. M. Gray and D. H. Male (eds), *Handbook of Snow: Principles, Processes, Management and Use* (Caldwell, NJ: Blackburn Press, 2004).

Discovery of ice ages: Louis Agassiz, *Études sur les glaciers* [Study on glaciers], 2 vols. (Neuchâtel: Lithographie H. Nicolet, 1840).

Observation of firn grains: C. S. Wright and R. E. Priestley, *British (Terra Nova) Antarctic Expedition, 1910–1913: Glaciology* (London: Harrison and Sons, 1922); H. Wilhelmsson Ahlmann, 'Review of *Glaciology*, by Wright & Priestley', *Geografiska Annaler* (journal of the Swedish Society for Anthropology and Geography), 7 (1925), pp. 155–62.

White ice: Albert Heim, *Poggendorffs Annalen der Physik und Chemie* [Poggendorff's Annals of Physics and Chemistry] (1870); Albert Heim, *Handbuch der Gletscherkunde* [Handbook of glaciology] (Stuttgart, 1885); Gerald Seligman, *Snow Structure and Ski Fields: Being an Account of Snow and Ice Forms Met With in Nature and a Study of Avalanches and Snowcraft* (London: MacMillan, 1936), p. 119.

Age of ice in Antarctica: J. Hager, 'New record: 4.6 million-year-old ice found in Antarctica', *Polar Journal*, 24 April 2024.

Theory of continental drift: Naomi Oreskes, *The Rejection of Continental Drift* (Oxford University Press, 1999). Among those who embraced Wegener's theory were several influential Nazis – see Eric Buffetaut, 'Continental drift under the Third Reich', *Endeavour*, 27:4 (2003), pp. 171–4.

Ice drilling in 1913: J. P. Koch and A. Wegener, 'Wissenschaftliche Ergebnisse der Dänischen Expedition nach Dronning Louises-Land und quer über das Inlandeis von Nordgrönland 1912–13' [Scientific results of the Danish expedition to Dronning Louise's Land and across the inland ice of North Greenland 1912–13], *Meddelelser om Grönland*, 75, (Köpenhamn, 1930); Mott T. Greene, *Alfred Wegener: Science, Exploration, and the Theory of Continental Drift* (Baltimore: Johns Hopkins University Press, 2015), p. 303.

Isotope dating and planetary time: Erik Isberg, *Planetary Timemaking: Paleoclimatology and the Temporalities of Environmental Knowledge, 1945–1990* (Stockholm: KTH, 2023), pp. 48–55 and ch. 3 (about Dansgaard).

New relevance of Little Ice Age: S. Þórarinsson, 'Tefrokronologiska studier på Island: Pjórsárdalur och dess förödelse' [Tephrochronological studies in Iceland: Pjórsárdalur and its devastation], *Geografiska Annaler*, 26 (1944), pp. 1–217; G. Utterström, 'Climatic fluctuations and population problems in early modern history', *Scandinavian Economic History Review*, 3 (1955), pp. 3–47; Robert W. Rix, *The Vanished Settlers of Greenland: In Search of a Legend and Its Legacy* (Cambridge University Press, 2023).

Back thousands of years: Richard B. Alley, *The Two-Mile Time Machine* (Princeton University Press, 2005).

Nakaya on ice cores: Ukichiro Nakaya (1957), 'Moon World in White', cited in Kenjiro Okazaki, 'What overflows', *Kaga Kuukan*, 20 April 2018, https://kagakuukan.org/eng/archive/what-overflows

Problem of glacial formation: Seligman, *Snow Structure and Ski Fields*.

Seligman Inlet: 'Gerald Seligman', *Wikipedia*; J. W. Glen, 'The Journal of Glaciology: Its origin and early history', *Journal of Glaciology*, 56:200 (2010), pp. 941–3.

Limits of glaciology: G. S. Callendar, 'Glacier fluctuations', *Quarterly Journal of the Royal Meteorological Society*, 70:305 (1944), p. 22 – a note from Ahlmann is cited in which he dismisses Callendar, which should be interpreted to mean that glacier research cannot be invoked in support of the carbon dioxide hypothesis; Sverker Sörlin, 'The anxieties of a science diplomat: field co-production of climate knowledge and the rise and fall of Hans Ahlmann's "Polar Warming"', *Osiris*, 26:1 (2011); R. Fleming and V. Jankovich (eds), *Revisiting Klima* (University of Chicago Press, 2011), p. 84.

Callendar's theory of climate change: James R. Fleming, *The Callendar Effect: The Life and Work of Guy Stewart Callendar* (Boston: American Meteorological Society, 2007).

Snow as a muffler of sound: M. Kessiby, 'Why does everything get so quiet after a snow fall?', Montréal Science Centre blog 24 Feb. 2023; M. Elischer, 'Snow science: silent snow', Michigan State University Extension, 21 Dec. 2016; 'Modes of sound wave propagation', Iowa State University, https://www.nde-ed.org/Physics/Sound/modepropagation.xhtml

Five centimetres of snow: University of Kentucky, 'The science behind snow's serenity', *Science Daily*, 21 Jan. 2016, www.sciencedaily.com/releases/2016/01/160121150503.htm

Snow and crime: R. Diary Ali and E. Swartling, 'Vädrets inverkan på utvalda kriminella handlingar inom Stockholms län: En studie baserad på linjär regression' [The impact of weather on selected criminal offences in Stockholm County: A linear regression study] (Stockholm: KTH, 2018). US data points in the same direction: 'Is there less crime when it snows?', B Safe Security, https://bsafealarms.com/uncategorized/is-there-less-crime-when-it-snows/

Influence of weather on crime: E. A . Burakowski et al., 'Future of winter in Northeastern North America: climate indicators portray warming and snow loss that will impact ecosystems and communities', *Northeastern Naturalist*, 28 (2022), pp. 180–207; N. J. Casson et al., 'Winter weather whiplash: impacts of meteorological events misaligned with natural and human systems in seasonally snow-covered regions', *Earth's Future*, 7:12 (2019), pp. 1434–50; T. Christopher Thomas et al., 'Weird winter weather in the Anthropocene: how volatile temperatures shape violent crime', *Journal of Criminal Justice*, 87 (2023), p. 102090.

Emmanuelle Charpentier in Umeå: Johanna Fredriksson, 'Emmanuelle Charpentier: "I really miss the atmosphere at Umeå University"', Umeå University, 14 Oct. 2020, https://www.umu.se/en/feature/historic-celebration-at-umea-university/

Blizzard disaster in the Anaris mountains: articles from the Swedish Alpine Club: https://fjallklubben.se/klubben/sfk-tipsar/92-anarisolyckan-40-ar; https://fjallklubben.se/klubben/sfk-tipsar/93-anarisolyckan-meteorologiska-orsaker

A sudden and violent snowstorm: https://www.smhi.se/kunskapsbanken/meteorologi/stormar-i-sverige/20240222/norden; Mats Ekdahl, *Snöns historia* [The history of snow] (Stockholm: Carlsson Bokförlag), pp. 228–9.

Eine Frau: Christiane Ritter (1938), *Eine Frau erlebt die Polarnacht* [A woman in the polar night] (London: Pushkin Press, 2019), pp. 90–9.

'spectacularly sublime': J. M. Giles, 'Conrad's avant-garde sublime: spectacular language, nature, and the Other in *Typhoon*', *Conradiana*, 47:3 (2015), pp. 163–210.

Swedish wind speed record: https://www.smhi.se/kunskapsbanken/meteor ologi/svenska-vindrekord

Sound of katabatic wind: Philip Samartzis, 'Katabatic winds inspire sound rendition of Antarctic experience', *Australian Antarctic Magazine*, 30 (2016).

Include the oceans in historiography: David Armitage, Alison O. Bashford and Sujit Sivasundaram (eds), *Oceanic Histories* (Cambridge University Press, 2017).

Eye-opening bestseller: Rachel Carson, *The Sea Around Us* (Oxford University Press, 1951).

Andrée alone in the Arctic snow: S. Sörlin, 'Inför dödens vita mörker: fältdag boken som konst och vetenskap' [Facing the white darkness of death: the field diary as art and science], in *Svenska fysikaliska-meteorologiska expeditionen till Spetsbergen juli 1882–september 1883: S. A. Andrées dagbok från deltagandet i det första internationella polaråret* [Swedish physical-meteorological expedition to Spitsbergen July 1882–September 1883: S. A. Andrée's diary from the participation in the first International Polar Year], ed. H. Jorikson (Grenna Museum, 2008).

SNOW GRIEF

Looking for insight: Judith Schalansky, *An Inventory of Losses*, tr. Jackie Smith (New Directions and MacLehose Press, 2020).

Isn't to say that grief is inherently healthy: F. Hedlund, 'Svår sorg ger sämre hälsa' [Severe grief makes health worse], *Medicinsk vetenskap* [*Medical Science*], Karolinska Institutet, 23 May 2023.

Animals feel sadness: Barbara J. King, *How Animals Grieve* (University of Chicago Press, 2013).

'Evolutionary thanatology': J. R. Anderson et al., 'Evolutionary thanatology', *Philosophical Transactions B*, 373:1754 (2018).

Friendship with life: E. Gullone, 'The biophilia hypothesis and life in the 21st century: increasing mental health or increasing pathology?', *Journal of Happiness Studies*, 1:3 (2000), pp. 293–321.

Samvettet: Sara Lidman, 'Före ordet' [Before the word], in *Och trädet svarade* [And the tree answered] (Stockholm: Bonniers, 1980), p. 11. The word *samvett* – literally, knowing and sensing, both at the same time – started as a misspelling, Sara Lidman points out, and became a remarkable neologism.

Greenlanders' fear of climate change: K. Minor et al., 'Greenlandic perspectives on climate change 2018–2019: results from a national survey' (University of Greenland et al., 2019).

Solastalgia in the Arctic: O. Michelin, '"Solastalgia": Arctic inhabitants overwhelmed by new form of climate grief', *Guardian* 15 Oct. 2020.

Concept of solastalgia: G. Albrecht et al., 'Solastalgia: The distress caused by environmental change', *Australian Psychology*, 15 (2007), suppl. 1, pp. 95–8; Albrecht, *Earth Emotions: New Words for a New World* (Ithaca, NY: Cornell University Press, 2019); Albrecht, 'Chronic environmental change: emerging "psychoterratic" syndromes', in I. Weissbecker (ed.), *Climate Change and Human Well-being: Global Challenges and Opportunities* (New York: Springer, 2011), pp. 43–56.

Uninhabitable places: L. J. Martin, 'Proving grounds: ecological fieldwork in the Pacific and the materialization of ecosystems', *Environmental History*, 23:3 (2018), pp. 567–92.

Earth's average temperature rising: 'June 2024 marks the 12th month of global temperatures at 1.5°C above pre-industrial levels', *Copernicus*, 10 July 2024.

'The earth is faster now': Legraaghaq, a tribal elder of the Yupik people, *c.*1912, cited in Igor Krupnik and Dyanna Jolly (eds), *The Earth Is Faster Now: Indigenous Observations of Arctic Environmental Change* (Fairbanks: Arctic Research Consortium of the United States, 2002), p. 7; Bathsheba Demuth, *Floating Coast: An Environmental History of the Bering Strait* (New York: Norton, 2019), p. 10.

Psychological effects of climate change: H. Comtesse et al., 'Ecological grief as a response to environmental change: a mental health risk or functional response?', *International Journal of Environmental Research on Public Health*, 16 (2021), p. 734; S. Braun, 'Warmer winters and vanishing snow breed climate grief', *Deutsche Welle*, 12 Jan. 2023.

New terms have come to the fore: A. Cunsolo et al., 'Ecological grief and anxiety: the start of a healthy response to climate change?', *Lancet Planetary Health*, 4 (2020), pp. e261–3; M. Ojala et al., 'Anxiety, worry, and grief in a time of environmental and climate crisis: a narrative review', *Annual Review of Environment and Resources* 46:1 (2021), pp. 35–58.

Words matter: N. Wormbs and M. Wolrath Söderberg, 'Knowledge, fear, and conscience: reasons to stop flying because of climate change', *Urban Planning*, 6:2 (2021), pp. 314–24.

Systematic smear campaigns: Naomi Oreskes and Erik M. Conway, *Merchants of Doubt: How a Handful of Scientists Obscured the Truth on Issues from Tobacco Smoke to Global Warming* (New York: Bloomsbury, 2011).

Winter grief: P. Pihkala, 'Climate grief: how we mourn a changing planet', BBC, 3 April 2020, https://www.bbc.co.uk/future/article/20200402-climate-grief-mourning-loss-due-to-climate-change

Easier for women than for men: K. Ekberg and V. Pressfeldt, 'A road to denial: climate change and neoliberal thought in Sweden, 1988–2000', *Contemporary European History*, 31:4 (2022), pp. 627–44; K. Ekberg and M. Hultman, 'A question of utter importance: the early history of climate change and energy policy in Sweden, 1974–1983', *Environment and History*, 29:3 (2023), pp. 399–421.

In the grip of rage: M. Ojala, 'How do young people deal with border tensions when making climate-friendly food choices? On the importance of critical emotional awareness for learning for social change', *Climate*, 10:1 (2022).

Terrafurie: G. Albrecht, 'Terrafurie = Earth Anger', *Psychoterratica*, 12 July 2017, https://glennaalbrecht.wordpress.com/2017/07/12 /terrafurie/

Losses other than snow: Marika Palmér Rivera and Lisa Pelling, *Livet som pågår här: Samtal om klimatet och omställningen* [The life that's happening here: Conversations about climate and the transition] (Stockholm: Atlas, 2024).

Helplessness in the face of snow anxiety: P. Pihkala, 'Eco-anxiety', in C. Parker Krieg and Reetta Toivanen (eds), *Situating Sustainability: A Handbook of Contexts and Concepts* (Helsinki University Press, 2021), pp. 119–34; P. Pihkala, 'Climate grief', op. cit.

Concern about climate change: Z. Provant et al., 'Who is killing the glaciers? From glacier funerals to glacier autopsies', *Edge Effects*, 3 Nov. 2023, https://edgeeffects.net/glacier-funerals/; J. C. Hu, 'The decade of attribution science', *Slate*, 19 Dec. 2019, https://slate.com/technology/2019/12/attribution-science-field-explosion-2010s-climate-change.html

COLD ACTIVISM

The Ice Palace: Tarjei Vesaas, *The Ice Palace*, tr. Elizabeth Rokkan (London: Peter Owen Publishers, 1966).

Unfortunate mountaineers: N. Singh Khadka, 'Mount Everest: Melting glaciers expose dead bodies', BBC, 21 March 2019, https://www.bbc.com/news/science-environment-47638436

As snowfields and glaciers melt: E. J. Dixon et al., 'The emergence of glacial archaeology', *Journal of Glacial Archaeology*, 1 (2014), pp. 1–9; M. C. Ceruti, 'Overview of the Inca frozen mummies from Mount Lullaillaco (Argentina)', *Journal of Glacial Archaeology*, 1 (2014), pp. 79–97.

Discarded plutonium: 'Nanda Devi Plutonium Mission', *Wikipedia*, https://en. wikipedia.org/wiki/Nanda_Devi_Plutonium_Mission

Ongoing research in Swedish mountains: K. Å. Aronsson, 'Fynd ur snölegan' [Discoveries in the snowbank], in Lotten Gustafsson-Reinius (ed.), *Arktiska spår: Natur och kultur i rörelse* [Arctic tracks: Nature and culture in motion] (Stockholm: Nordiska Museets Förlag), pp. 190–1.

Atlas of Argentina's glaciers: J. Höglund Hellgren, 'Negotiating governable objects: glaciers in Argentina', in Klaus Dodds and Sverker Sörlin (eds), *Ice Humanities: Living, Working, and Thinking in a Melting World* (Manchester University Press, 2022).

Argentinian legal protection for its glaciers: Jorge D. Taillant, *Glaciers: The Politics of Ice* (New York: Oxford University Press, 2015).

Project in the Andes: D. Collyns, 'Can painting a mountain restore a glacier?', BBC, 17 June 2010, https://www.bbc.com/news/10333304; C. Atkinson, 'Painting rock while glaciers melt', *Pique News Magazine*, 23 March 2012, https://www.piquenewsmagazine.com/whistler-news/painting-rock-while-gla-ciers-melt-2490802

World Heritage status, Ilulissat: UNESCO World Heritage Convention, 'Ilulissat Icefjord', https://whc.unesco.org/en/list/1149/; N. Mikkelsen and T. Ingerslev (eds), 'Nomination of the Ilulissat Icefjord for inclusion in the World Heritage List (2002)', https://whc.unesco.org/uploads/nominations/1149.pdf

Technological escalation: D. MacAyeal et al., 'Glacial Climate Intervention: A Research Vision', University of Chicago white paper, 13 June 2024, https://climate.uchicago.edu/wp-content/uploads/2024/10/Glacial-Climate-Intervention_A-Research-Vision.pdf

'Defence campaigns': M. Carey, 'The history of ice: how glaciers became an endangered species', *Environmental History*, 12:3 (2007); M. Carey et al., 'Glacier protection campaigns: What do they really save?', in Dodds and Sörlin, *Ice Humanities*, pp. 89–109, op cit.

Ancient techniques for preserving ice and snow: J. Rees, 'The many ways that water froze: A taxonomy of ice in nineteenth- and early twentieth-century America', in Dodds and Sörlin, *Ice Humanities*, op cit.

Artificial cold: Kostas Gavroglu (ed.), *History of Artificial Cold, Scientific, Technological and Cultural Issues* (Dordrecht: Springer, 2014); Shane Hamilton, *Trucking Country: The Road to America's Wal-Mart Economy* (Princeton University Press, 2008); Terje Finstad et al., *Varme visjoner og frosne fremskritt: Om fryseteknologi i Norge, ca. 1920–1965* [Warm visions and frozen progress:

About freezing technology in Norway, ca. 1920–1965] (Trondheim: NTNU, 2011); Nicola Twilley, *Frostbite: How Refrigeration Changed Our Food, Our Planet, and Ourselves* (New York: Penguin, 2024), pp. 25–62.

New techniques for glacier protection: Carey et al., 'Glacier protection campaigns', op cit.

Hundreds of ski resorts have now closed: V. Mitterwallner et al., 'Global reduction of snow cover in ski areas under climate change', *PLoS ONE*, 19:3 (2024), pp. e0299735, https://doi.org/10.1371/journal.pone.0299735; M. Beeck, 'Här har snöbrist stängt 300 skidanläggningar' [Lack of snow has closed 300 ski resorts here], *Dagens PS*, 27 Dec. 2022; https://www.dagensps.se/weekend/resor/har-har-snobrist-stangt-300-skidanlaggningar/

Numbering in the hundreds by 1914: M. Wedekind, 'Mountain grand hotels at the fin de siècle: sites, gazes, and environments', *Journal of the Austrian Association for American Studies*, 2:2 (2021), pp. 163–87.

Monuments to time and snow gone by: Wally Koval, *Accidentally Wes Anderson* (London: Orion, 2020).

Chewang Norphel: H. Shrager, '"Ice Man" vs. Global Warming', *Time Magazine*, 25 Feb. 2008.

2019 international study: 'Water, ice, society, and ecosystems in the Hindu Kush Himalaya: an outlook', ICIMOD, https://hkh.icimod.org/hi-wise/hi-wise-report/. The report is produced by the International Centre for Integrated Mountain Development (ICIMOD), from their base in Katmandu. ICIMOD also coordinates the Save Our Snow campaign – see http://www.icimod.org/saveoursnow

Water availability has declined: C. Clason et al., 'Contribution of glaciers to water, energy and food security in mountain regions: current perspectives and future priorities', *Annals of Glaciology*, 63:87–89 (2022), pp. 73–8.

Quality of meltwater and toxic substances: Andreas Karlsson, *Vatten: En historia om människor och civilisationer* [Water: a story of people and civilizations] (Lund: Historiska Media, 2021); D. Drage, 'Forever chemicals in ski wax are being spread on snowy slopes', *The Conversation*, 12 Feb. 2024, https://theconversation.com/forever-chemicals-in-ski-wax-are-being-spread-on-snowy-slopes-222095; Ian. T. Cousins et al., 'Outside the safe operating space of a new planetary boundary for per- and polyfluoroalkyl substances (PFAS)', *Environmental Science & Technology*, 56:16 (2022), pp. 11172–9; J. McKoy, 'Skiers and snowboarders face high risk of exposure to PFAS', Boston University School of Public Health, 10 March 2023, https://www.bu.edu/sph/news/articles/2023/skiers-and-snowboarders-face-high-risk-of-exposure-to-pfas/

Rhetoric overshadows results: Carey et al., 'Glacier protection campaigns'; R. E. Carver and F. S. Tweed, 'Cover the ice or ski on grass?', *Geography*, 106:3 (2021), pp. 116–27.

Art efforts, in tandem with engineering projects: L. Palmer, 'Olafur Eliasson responds to Paris Summit with a Doomsday clock made of glacial ice', *Artnet*, 3 Dec. 2015, https://news.artnet.com/art-world/ice-watch-olafur-eliasson-climate-summit-384704

Dying monument: Rebecca Solnit, 'Power in Paris', *Harper's Magazine*, Dec. 2015.

Summit of Kebnekaise: Bigert & Bergström, '*The Freeze, 2015: Rescue Blanket for Kebnekaise*', https://bigertbergstrom.com/works/the-freeze

Choices that societies now have to make: S. Sörlin, 'Reverse geoengineering: on *The Weather War* exhibition, Shanghai, and the performance climate activism by Bigert & Bergström', in *The Weather War* (Shanghai Minsheng Art Museum, 2017), pp. 40–7.

'Truth spot': Thomas Gieryn, *Truth Spots: How Places Make People Believe* (The University of Chicago Press, 2018).

Eyes that first saw: J. Esmark, 'Bidrag til vor jordklodes historie' [Contributions to the history of our planet], *Magazin for Naturvidenskaberne*, 2:1 (1824), pp. 28–49; G. Hestmark, 'Jens Esmark's mountain glacier traverse 1823 – the key to his discovery of Ice Ages', *Boreas*, 47 (2018), pp. 1–10; Tobias Krüger, *Discovering the Ice Ages: International Reception and Consequences for a Historical Understanding of Climate* (Leiden: Brill, 2013); G. Hestmark, 'How Jens Esmark discovered the Ice Age in 1823 – serendipity and a prepared mind', *ResearchGate*, Oct. 2023; https://www.researchgate.net/publication/3745012 17_Anniversaries_Discovery_of_the_Ice_Age_200_years_ago_HOW_JENS_ ESMARK_DISCOVERED_THE_ICE_AGE_IN_1823_-_SERENDIPITY_AND_ A_PREPARED_MIND

Sweden greatest snow depths: L. Wern, 'Snödjup i Sverige 1904/05–2013/14' [Snow depth in Sweden 1904/05–2013/14], Swedish Meteorological and Hydrological Institute, report 158 (2015).

Snowfall in US states: National Snow and Ice Data Center, 'Snow', https://nsidc.org/learn/parts-cryosphere/snow/science-snow

Chacaltaya; Whistler: National Snow and Ice Data Center, 'Snow – why it matters: impact of climate change on snow', https://nsidc.org/learn/parts-cryosphere/snow/why-snow-matters#anchor-impact-of-climate-change-on-snow

Dismantled cable car: Marie-Louise Kristola, 'Franska skidliftar plockas ner när temperaturerna stiger' [French ski lifts are being taken down as temperatures rise], Sveriges Radio [Swedish Radio], 8 Feb. 2024, https://sver-igesradio.se/artikel/franska-skidliftar-forsvinner-nar-temperaturerna-stiger

New cable car: 'Chamonix: new Telecabine between Montenvers & the Mer de Glace', 15 Feb. 2024, https://www.chamonix.net/english/news/chamonix-new-telecabine-montenvers-mer-de-glace

An earth of 'discontinuities': Charlotte Wrigley, *Earth, Ice, Bone, Blood: Permafrost and Extinction in the Russian Arctic* (Minneapolis: University of Minnesota Press, 2023).

Melting is accelerating: M. Langer et al., 'The evolution of Arctic permafrost over the last three centuries from ensemble simulations with the Cryo-GridLite permafrost model', *Cryosphere*, 18:1 (2024), pp. 363–85.

THE CRYSTAL SEEKERS

Predicting the future of our planet: T. Bartels-Rausch, 'Ten things we need to know about ice and snow', *Nature*, 494:7435 (2013), pp. 27–29.

We take on new chrono-geographical narratives: Anders Ekström and Staffan Bergwik (eds), *Times of History, Times of Nature: Temporalization and the Limits of Modern Knowledge* (New York: Berghahn, 2022).

Authority on snow crystals: Kenneth G. Libbrecht, *Snow Crystals: A Case Study in Spontaneous Structure Formation* (Princeton University Press, 2022).

Han Yin citation: Ibid., p. 8.

Epistemic community: Peter M. Haas, 'Introduction: epistemic communities and international policy coordination', *International Organization*, 46:1 (1992), pp. 1–35.

How little was understood: Libbrecht, *Snow Crystals*, op. cit., p. 1.

It's about transformation: Thomas Seiler, *Snøens formler: Et naturfenomen i litteraturen* [The formulas of snow: a natural phenomenon in literature] (Oslo: Cappelen Damm, 2023).

Another classical writer: Pliny the Elder, *Natural History,* tr. J. Bostock, vol. 2, ch. 61 (London: Henry G. Bohn, 1855), pp. 90–91.

Greek geographer Strabo: Strabo (probably 16 CE), *Geographica* IV, book 6.

'snow is like feathers': Herodotus, *The Histories*, tr. George Rawlinson, book 4, paragraph 31 (London: John Murray, 1859), p. 26.

An amusing comment: Epictetus, *Discourses and Selected Writings*, tr. and ed. Robert Dobbin (London: Penguin Classics, 2008).

Snow rare in books of antiquity: Kate Gilhuly and Nancy Worman (eds),

Space, Place and Landscape in Ancient Greek Literature and Culture (Cambridge: Cambridge University Press, 2014).

'during a very heavy snow-storm': Josephus, *The Jewish War*, book 1, ch. 16 (London: Penguin, 1981).

William Turner's Hannibal painting: Joseph Mallord William Turner, *Snow Storm: Hannibal and his Army Crossing the Alps*, https://www.tate.org.uk/art/art-works/turner-snow-storm-hannibal-and-his-army-crossing-the-alps-n00490

Argument over Hannibal's route: Philip Ball, 'The truth about Hannibal's route across the Alps', *Guardian* 3 April 2016, https://www.theguardian.com/science/2016/apr/03/where-muck-hannibals-elephants-alps-italy-bill-mahaney-york-university-toronto

'the petals of roses falling like snow': Lucretius, *On the Nature of Things*, Gutenburg Project, trans. William Ellery Leonard, https://www.gutenberg.org/files/785/785.txt

Visions of Paradise: Mary B. Campbell, *The Witness and the Other World: Exotic European Travel Writing, 400–1600* (Ithaca, NY: Cornell University Press, 1988).

Whip-smart Albertus Magnus: Albertus Magnus, *De Meteoris* (*c.*1260), ch. 1, 10 (re. Aristotle, *Meteorologica* vol. 1, ch. 11, p. 19). One of the most developed commentators on snow in the Middle Ages. In *De Meteoris*, Magnus writes: 'For snow must necessarily be white, since it is itself composed of parts of transparent material with clotted air dispersed among them'; Magnus also mentions snow several times in his work *On the Causes of the Properties of the Elements*, tr. Dorothy Wyckoff (Oxford: Clarendon Press, 1967).

Albertus Magnus observation popularized: J. Needham and L. Gwei-Djen, 'The earliest snow crystal observations', *Weather*, 16:10 (1961) pp. 10, 319–27, https://rmets.onlinelibrary.wiley.com/doi/10.1002/j.1477-8696.1961.tb02589.x

China ahead of Europe: S. Kink, 'The explanations of snow in the *Taixi shuifa* 西水法 泰西水法 (Hydromethods of the Great West, 1612) and Their Reception beyond the Ming–Qing Transition', *Monumenta Serica*, 70:1 (2022), pp. 165–207. Later, however, the order was reversed in what has been called the 'Great Divergence' of the early modern period, when Chinese scholars proved reluctant to take up the then growing snow research in Europe.

Snow in *The Divine Comedy*: Dante Alighieri, *The Divine Comedy*, tr. Allen Mandelbaum (London: Everyman's Library, 1995), *Inferno*: canto VI, ll. 10–11, canto XIV, l. 30; *Purgatorio*: canto XXI, ll. 46–7; *Paradiso*: canto II, ll. 106–11, canto XXXIII, ll. 61–4.

In the very cold 1560s: Elke Oberthaler, et al. (eds), *Bruegel: The Hand of the Master* (Wien: Kunsthistorisches Museum, 2018).

Little Ice Age: F. E. Matthes, 'Report of Committee on Glaciers, April 1939', *Transactions of the American Geophysical Union*, 20:4 (1939), pp. 518–23; Brian Fagan, *The Little Ice Age: How Climate Made History, 1300–1850* (New York: Basic Books, 2000).

Ice ages before that: 'Timeline of glaciation', Wikipedia, https://en.wikipedia. org/wiki/Timeline_of_glaciation

SNOW CURES MADNESS

Peoples of the Altai mountains: K. Krichko, 'China's Stone Age skiers and history's harsh lessons', *New York Times*, 19 April 2017, https://www.nytimes. com/2017/04/19/sports/skiing/skiing-china-cave-paintings.html

Dobrowolski's monumental book: Antoni B. Dobrowolski, *Historia naturalna lodu* [Natural History of Ice] (Warsaw: Kasa im Mianowskiego, 1923).

Tears and weakness: G. Seligman, 'Cryology', *Journal of Glaciology*, 1:1 (1947), p. 35.

Dobrowolski was devastated: Roger G. Barry et al., 'A. B. Dobrowolski – the first cryospheric scientist – and the subsequent development of cryospheric science', *History of Geo and Space Sciences* 2 (2011), pp. 75–9.

Silencing all doubters: Ian Allison et al., 'IACS: past, present, and future of the International Association of Cryospheric Sciences', *History of Geo and Space Sciences* 10 (2019), pp. 97–107.

A popular idea: N. C. Knight, 'No two alike?', *Bulletin of the American Meteorological Society*, 69:5 (1988), p. 496.

'Six-Cornered Snowflake': Johannes Kepler (1611), *On the Six-Cornered Snowflake*, trans. L. L. Whyte (Oxford University Press, 1966), http://www. joostwitte.nl/M_Galilei/Johannes_kepler_snowflake.pdf

A New Year's present: Cecil Schneer, 'Kepler's New Year's gift of a snowflake', *Isis*, 51:4 (1960), pp. 531–45; F. Winckel, 'Perhaps the most famous New Year's gift in science', *Doing History in Public*, 23 Dec. 2023, https://doinghistoryin-public.org/2023/12/23/23-perhaps-the-most-famous-new-years-gift-in-science/

'De Nihilo': P. White, 'The poetics of nothing: Jean Passerat's "De Nihilo" and its legacy', *Erudition and the Republic of Letters*, 5:3 (2020), pp. 237–73. Thanks to Floris Winckel for this reference.

Kepler not first to take interest: Kenneth G. Libbrecht, *Snow Crystals: A Case Study in Spontaneous Structure Formation* (Princeton University Press, 2022), ch. 1 – includes summaries of the long-standing canon of names in snow star research, from Aristotle to the present day; B. J. Ford, 'The hidden secrets of snowflakes', *Microscope*, 62:4 (2014), pp. 171–81.

'the particles of ice': René Descartes, *Discours de la methode* (Leiden: De l'imprimerie de I. Maire, 1637), p. 232; F. C. Frank, 'Descartes' observations on the Amsterdam snowfalls of 4, 5, 6 and 9 February 1634', *Journal of Glaciology*, 13:69 (1974), pp. 535–9.

Bartholin brothers' work: Thomas Bartholin, *De nivis usu medico observationes variae* [Various observations on the medical use of snow] (Copenhagen: Matthias Godicchen, 1661).

Microscopic representation of snow: Robert Hooke, *Micrographia* (London: J. Martyn and J. Allerstry, 1665).

Suggested Hooke stole images: B. J. Ford, 'The hidden secrets of snowflakes', op. cit.; B. J. Ford, 'Robert Hooke's *Micrographia*' (1998), https://www.brianj-ford.com/a98-hooke.htm

Boyle and Cavendish: C. Rosengren, 'Touching the cold in the Little Ice Age: Reason and fancy in Robert Boyle's and Margaret Cavendish's writings on northern cold', *Lychnos*, 87 (2022), pp. 155–74.

Wilcke's snowscape experiments: J. C. Wilcke, 'Rön och tankar om snö-figurers skiljaktighet' [Evidence and thinking on the discordance of snow figures], in *Kungl. Vetenskapsakademiens handlingar* [Royal Academy of Science documents], 22 (1761), pp. 1–19; 'Nya rön om vattnets frysning til snölike is-figurer' [New evidence on water freezing into snow-like ice shapes], *KVAH*, 30 (1769), pp. 90–111; 'Om snöns kyla vid smältningen' [On the coldness of snow as it melts], *KVAH*, 33 (1772), pp. 97–120.

WHITE MAGIC

Scoresby's Arctic research on snow stars: William Scoresby, *An Account of the Arctic Regions with a History and Description of the Northern Whale-fishery*, vol. 2 (Edinburgh: A. Constable and Co., 1820).

Martens' classification: Friedrich Martens, *Spitzbergische oder Groenländische Reise Beschreibung gethan im Jahr 1671* [Spitzbergen or Greenlandic journey description made in the year 1671] (Hamburg: Gottfried Schultzens, 1675).

'pedagogical enlightenment': Marleen de Vries, 'Literature of the Enlightenment, 1700–1800', in Theo Hermans (ed.), *Literary History of the Low Countries* (London: Boydell and Brewer, 2009), pp. 293–367.

Minister's wife in Portland: Frances E. Chickering, *Cloud Crystals: A Snow Flake Album* (New York: D. Appleton & Company, 1864).

SELECTED REFERENCES

'If you keep on believing': *Cinderella*, animated musical film (Walt Disney Productions, 1950), lyrics by Mack David, Jerry Livingston and Al Hoffman.

LETTERS FROM HEAVEN

Fujiko Nakaya's 'fog sculptures': F. Winckel, 'Frost flowers and fog sculptures', *HUBE Magazine* 1:1 (2023), pp. 314–31.

Precious gem heirlooms: Thomas Mann, *The Magic Mountain* (1924), tr. H. T. Lowe-Porter (London: Secker & Warburg, 3rd edn, 1961), p. 480.

Snow over evil: Orhan Pamuk, *Snow*, tr. M. Freely (New York: Alfred A. Knopf, 2011).

Russian A. A. Sigson: S. A. Sokratov, 'The Russian contribution to snow science', *ICE* (2013), pp. 4–9.

Nordenskiöld also contributed: G. Nordenskiöld, 'The photography of snow flakes', *Photographic Times*, 26 (1895); 'Communication préliminaire sur une étude des cristaux de neige' [Preliminary communication on a study of snow crystals], *Bulletin de la Société Francaise de minéralogie* 16 (1893), pp. 59–74; W. Odelberg, 'Gustaf E. A. Nordenskiöld', in *Svenskt biografiskt lexikon* [Dictionary of Swedish National Biography], https://sok.riksarkivet.se/SBL/Start.aspx?lang=en.

Bentley's published work: 'The magic beauty of snow and dew', *National Geographic*, 43:1 (1923), p. 103; 'Snow', *Encyclopaedia Britannica*, 20 (1936), pp. 854–6.

Bentley's major work: Wilson A. Bentley and William J. Humphreys, *Snow Crystals* (New York: McGraw-Hill, 1931).

Volcanic eruptions and climate: W. J. Humphreys, 'Volcanic dust as a factor in the production of climatic changes', *Journal of the Washington Academy of Sciences*, 3:13 (1913), pp. 365–71.

International Commission on Snow: U. Radok, 'The International Commission on Snow and Ice (ICSI) and its precursors, 1894–1994', *Hydrological Sciences Journal*, 42:2 (1997), pp. 131–40; ICSI founder J. E. Church also invented an instrument for measuring the water content of snow cover – see B. Wright, 'University professor's invention from over a century ago continues to be impactful today', *Nevada Today*, 17 March 2021.

'I sought snow': R. Butterfield, 'Nevada's fantastic snow man', *Saturday Evening Post*, 222:31 (1950), pp. 25–93.

Nakaya the physicist: Akira Higashi, 'Ukichiro Nakaya – 1900–1962', *Journal of Glaciology*, 4:33 (1962), pp. 378–80.

'the simplest and most primary problem': Ukichiro Nakaya, *Snow Crystals: Natural and Artificial* (Cambridge, Mass: Harvard University Press, 1954), p. vi.

Reviews of Nakaya's book: See K. F. Mather, *American Scientist*, 42:3 (1954), p. 508; J. L. Greenstein, *Scientific American*, 191:3 (1954), pp. 162–3; K. L. S. Gunn, *Scientific Monthly*, 80:1 (1955), p. 59; H. B. Nichols, *Science*, 120:3123 (1954), p. 755; P. Anker and S. Sörlin, 'Ukichiro Nakaya's Sense of Snow', in J. H. Engqvist and M. Hultman (eds), *Letters Sent from Heaven: Frozen and Vaporized Water: Ukichiro Nakaya and Fujiko Nakaya's Science and Art* (Oslo: OK Book, 2022), pp. 125–32.

Rosetta Stone for snowflakes: Kenneth G. Libbrecht, *Snow Crystals: A Case Study in Spontaneous Structure Formation* (Princeton University Press, 2022), p. 15.

STRATEGIC SNOW

Could Greenland melt away?: E. Storgaard, 'Alfred Wegeners Grønlandsekspedition 1929–1931', *Geografisk Tidsskrift* [Geographical journal], 35:4 (1932), pp. 198–214.

Station Eismitte: Janet Martin-Nielsen, *Eismitte in the Scientific Imagination: Knowledge and Politics at the Center of Greenland* (New York: Palgrave, 2017); Mott T. Greene, *Alfred Wegener: Science, Exploration, and the Theory of Continental Drift* (Baltimore: Johns Hopkins University Press, 2015).

Sorge's Law of Snow Densification: H. Bader, 'Sorge's law of densification of snow and firn on high polar glaciers', *Journal of Glaciology*, 15 (1954), pp. 319–23.

Swede Valter Schytt: Anna Schytt, *Med känsla för is: Om polarforskaren Valter Schytt och gåtorna hans Antarktisexpedition bidrog till att lösa* [With a sense for ice: Polar explorer Valter Schytt and the mysteries his Antarctic expedition helped to solve] (Stockholm: Fri Tanke, 2018), pp. 91–9.

The project's mission statement: D. Achermann, 'Vertical glaciology: The second discovery of the third dimension in climate research', *Centaurus* (2020), pp. 1–24.

Keeling launches his measuring instrument: S. Sörlin, 'The anxieties of a science diplomat: Field co-production of climate knowledge and the rise and fall of Hans Ahlmann's "Polar Warming"', *Osiris*, 26:1 (2011).

Science fiction-like research: Dian Olson Belanger, *Deep Freeze: The United States, the International Geophysical Year, and the Origins of Antarctica's Age of Science* (Boulder, CO: University Press of Colorado, 2006).

Political relevance of the greenhouse effect: U. Nakaya, 'Moon World in White' (1957), cited in Kenjiro Okazaki, 'What overflows', *Kaga Kuukan*, 20 April 2018, https://kagakuukan.org/eng/archive/what-overflows

Report from Mauna Loa: Ukichiro Nakaya, Juji Sugaya and Mikio Shoda, *Report of the Mauna Loa Expedition in the winter of 1956–57*, https://eprints.lib.hokudai.ac.jp/dspace/bitstream/2115/34229/1/5_P1-36.pdf

'danger of being submerged': Gorow Wakahama, 'From Snow and Ice World: Ukichiro Nakaya, a pioneer who saw the future of the global environment', in *Conversations with Snow and Ice* (Riga: Natural History Museum of Latvia, 2005), p. 12.

Soviet strong focus on permafrost and sea ice: Pey-Yi Chu, *The Life of Permafrost: A History of Frozen Earth in Russian and Soviet Science* (Toronto: University of Toronto Press, 2020); K. Doose, 'Pey-Yi Chu, The Life of Permafrost' review, *Cahiers du monde russe* 62:4 (2021), 740–43; J. Lajus and S. Sörlin, 'Melting the glacial curtain: The politics of Scandinavian–Soviet networks in the geophysical field sciences between two polar years, 1932/33–1957/58', *Journal of Historical Geography*, 44 (2014), pp. 44–59.

Analysing glacier movements: M. F. Perutz and G. Seligman, 'A crystallographic investigation of glacier structure and the mechanism of glacier flow', *Proceedings of the Royal Society of London, series A: Mathematical and Physical Sciences*, 172:950 (1939), pp. 339–40.

'A scientific Pearl Harbor': J. Martin-Nielsen, '"An orgy of hypothesizing": the construction of glaciological knowledge in Cold War America', in Julia Herzberg et al. (eds), *Ice and Snow in the Cold War: Histories of Extreme Climatic Environments* (New York: Berghahn Books, 2018), pp. 69–88.

A minimum of landmarks: C. Aporta et al., 'Pan Inuit Trails', 2014, https://paninuittrails.org/index.html?module=module.about

Stefánsson *persona non grata*: Gísli Pálsson, *Travelling Passions: The Hidden Life of Vilhjalmur Stefansson*, tr. K. Kunz (Winnipeg: University of Manitoba Press, 2005); David H. Price, *Threatening Anthropology: McCarthyism and the FBI's Surveillance of Activist Anthropologists* (Durham, NC and London: Duke University Press, 2004).

DOWN-TO-EARTH UTOPIA

Inuit words for snow: I. Krupnik et al., 'Franz Boas and Inuktitut terminology for ice and snow: From the emergence of the field to the "great eskimo vocabulary hoax"', in I. Krupnik et al. (eds), *SIKU: Knowing Our Ice* (Dordrecht: Springer, 2010); O. H. Magga, 'Diversity in Saami terminology for reindeer, snow, and ice', *International Social Science Journal*, 58:187 (2006), pp. 25–34.

Debate about words for snow: 'Inuktitut words for snow and ice', *Canadian Encyclopedia*, 9 July 2015; John L. Steckley, *White Lies About the Inuit*

(Peterborough, Ontario: Broadview Press, 2008); D. Robson, 'There really are 50 Eskimo words for "snow"', *Washington Post*, 14 Jan. 2013.

Poem about winter and existence: Lars Gustafsson, 'Världens tystnad före Bach' [The silence of the world before Bach] poem, tr. P. Martin, https://inwardboundpoetry.blogspot.com/2007/05/409-silence-of-world-before-bach-lars.html

The fox hunts: Ruth Kirk (1977), *Snow* (Seattle and London: University of Washington, 1998), p. 259.

The subnival: A. Froster, 'Livet i den hemliga världen under snön' [Life in the secret world under the snow], *Sveriges Natur* [Sweden's nature], 1 (2018).

Névé: 'Névé', Wikipedia, https://en.wikipedia.org/wiki/Névé

Firn: 'Firn', Wikipedia, https://sv.wikipedia.org/wiki/Firn

Open ocean around the North Pole: J. Lajus and S. Sörlin, 'An ice free Arctic Sea? The science of sea ice and its interests', in M. Christensen et al. (eds), *Media and Arctic Climate Politics* (New York: Palgrave, 2013), pp. 70–92; Michael Bravo, *North Pole: Nature and Culture* (London: Reaktion Books, 2019).

Franklin expedition: Edward Belcher, *The Last of the Arctic Voyages: In Search of Sir John Franklin, During the Years 1852–54, vol. 1* (Cambridge University Press, 2011).

Tyndall's experiments: J. Tyndall, 'On the absorption and radiation of heat by gases and vapours, and on the physical connexion of radiation, absorption, and conduction', *Philosophical Magazine and Journal of Science*, series 4, 22:146 (1861), pp. 169–94, 273–85.

Tyndall's research: B. Hevly, 'The heroic science of glacier motion', *Osiris*, 11 (1996), pp. 66–86.

Science and conquest: Ian Cameron, *To the Farthest Ends of the Earth: The History of the Royal Geographical Society 1830–1980* (London: Royal Geographical Society, 1980).

Achievement-oriented norms: D. Inkpen, 'Ever higher: The mountain cryosphere', in Klaus Dodds and Sverker Sörlin (eds), *Ice Humanities: Living, Working, and Thinking in a Melting World* (Manchester University Press, 2022), pp. 72–88.

Extreme and cold environments: Vanessa Heggie, *Higher and Colder: A History of Extreme Physiology and Exploration* (University of Chicago Press, 2019).

Sacrificing oneself for 'truth': Rebecca M. Herzig, *Suffering for Science* (New Brunswick, NJ: Rutgers University Press, 2005).

Protect Our Winters: https://protectourwinters.org/about-pow/ [accessed 21 April 2025]; https://protectourwinters.org/news/climate-advocates-guidebook/

Rosqvist–Inga collaboration: G. C. Rosqvist et al., 'Impacts of climate warming on reindeer herding require new land-use strategies', *Ambio*, 51 (2022), pp. 1247–62.

Influence public debate: D. Keyton, 'Reindeer starving as climate change affects Swedish Arctic, covering it in ice', Associated Press, 10 Dec. 2019, https://globalnews.ca/news/6277748/swedish-arctic-reindeer-starving/

Alliances around resource exploitation: Sverker Sörlin (1988), *Framtidslandet* [The land of the future] (Göteborg: Teg Publishing, 2023); Sverker Sörlin (ed.), *Resource Extraction and Arctic Communities* (Cambridge University Press, 2023).

Snow is a transitional substance: Robert Sharp, *Living Ice: Understanding Glaciers and Glaciation* (Cambridge University Press, 1988).

Viscous cryosphere: C. Simonetti, 'Viscosity in matter, life and sociality: The case of glacial ice', *Theory, Culture and Society*, 39:2 (2022), pp. 111–30.

Ecological properties of snow: W. O. Pruitt, Jr., 'Animals in the snow', *Scientific American*, 202:1 (1960), pp. 60–8; 'Snow as a factor in the wintering ecology of the barren ground caribou', *Arctic*, 12:3 (1959), pp. 159–79.

Concept of *qali*: W. Pruitt, 'Qali, a taiga snow formation of ecological importance', *Ecology*, 39:1 (1958), pp. 169–72.

Importance of snow cover for animals: Aleksandr Nikolajevitj Formozov, *Snow Cover as an Integral Factor of the Environment and Its Importance in the Ecology of Mammals and Birds*, tr. W. Prychodko and W. O. Pruitt (Edmonton: Boreal Institute for Northern Studies, University of Alberta, 1966).

Major breakthrough: C.-C. Coulianos, 'Djur och mikroklimat' [Animals and microclimate], *Zoologisk Revy* [Zoological review] 24 (1962), pp. 58–70; R. Geiger, *Das Klima der bodennahen Luftschicht* [The climate of the near-ground air layer] (Braunschweig: Friedr. Vieweg und Sohn, 1961).

They buried pit traps: Correspondence with Professor Kjell Danell, Swedish University of Agricultural Sciences, April 2024.

NIVEA

On the concept of 'outdoors' in Norwegian culture: Nina Witoszek, *Norske naturmytologier* [Norwegian nature myths] (Oslo: Pax Forlag, 1998); Peder Anker, *Livet er best ute: Friluftslivets historie og filosofi* [Life is best outdoors: The history and philosophy of outdoor recreation] (Oslo: Kagge, 2022).

Nansen looks feral: 'Portrett av Fridtjof Nansen, 1896' [Portrait of Fridtjof Nansen, 1896], Wikimeida Commons, https://commons.wikimedia.org/wiki/File:Portrett_av_Fridtjof_Nansen,_1896_(4583612654).jpg

Andrée's pedometer: 'Andrées sista läger påträffat av norsk arktisk expedition' [Andrée's last camp found by Norwegian Arctic expedition], *Svenska Dagbladet*, 23 Aug. 1930.

Evolution of the Beiersdorf Group: G. Jones and C. Lubinski, 'Managing political risk in global business: Beiersdorf 1914–1990', *Enterprise and Society* 13:1 (2012), pp. 85–119.

*Weiss in blau***:** 'NIVEA White in Blue 1936', Beiersdorf, *YouTube*, 20 Nov. 2020, https://www.youtube.com/watch?v=TeRupRdiNzM

*Katharine***:** 'NIVEA Historie: Deutschland TV-Spot "Katharine" (1938)', NIVEA Deutschland, *YouTube*, 29 Aug. 2012, https://www.youtube.com/watch?v=h68_GLe9j9M

Alpine sports started to take shape: 'Det var så det började' [That was how it began], *Alpin skidsport* [Alpine ski sports], 1972, pp. 102–7, https://www.ollerimfors.se/artiklar/Alpin_skidsport_1972.pdf; J. Schneider, 'Österrikiska undervisningsministeriets "Skidpedagogium i Tirolen"' [Austrian Ministry of Education 'ski school' in Tyrol], in *På skidor* [On skis], pp. 287–92.

Fascism's 'aestheticization': Walter Benjamin, 'The work of art in the age of mechanical reproduction', *Illuminations*, ed. Hannah Arendt, tr. Harry Zohn (New York: Schocken Books, 1969); Arne Ruth and Ingemar Karlsson, *Samhället som teater: Estetik och politik i Tredje riket* [Society as Theatre: Aesthetics and Politics in the Third Reich] (Stockholm: Liber, 1984), p. 14.

Winter Olympics: 'Games at Garmisch', *Time Magazine*, 17 Feb. 1936.

Porla Prize for Leni Riefenstahl: 'Schwedisches Porla-Preis für Leni Riefenstahl', *Völkischer Beobachter*, 8 June 1938.

Origin of International Winter Sports Week: S. Sörlin, 'Den snirkliga vägen till sportlovet' [The winding road to the sports holiday], *Forskning och framsteg* [Research and progress] 50:3 (2016), pp. 18–19; 'Filolog i sportkostym: Rektor Carl Svedelius' [Philologist in a sports suit: principal Carl Svedelius], in R. Ambjörnsson and S. Sörlin (eds), *Obemärkta: Det dagliga livets idéer* [Unnoticed: The ideas of daily life] (Stockholm: Carlssons, 1995), pp. 89–110.

1941 World Ski Championships: Leif Yttergren, *I och ur spår!: En studie om konflikter och hjältar i svensk skidsport under 1900-talet* [On and off the track!: A study of conflicts and heroes in Swedish skiing during the 1900s] (Lund: KFS [Swedish Ski Federation], 2006); Yttergren, 'Norway, skiing and Swede hate: Some reflections on the World Ski Championship in Cortina d'Ampezzo and its consequences', in H. Roiko-Jokela and P. Pöyhönen (eds), *The Many Faces of Snow Sports* (Jyväskylä University Press, 2017), pp. 226–40.

Göring's trophy hunt: Simon Schama, *Landscape and Memory* (London: HarperCollins Publishers, 1996).

'To Aryan blood': Viktor Rydberg, 'Himlens blå' [The blue of the sky] (poem, 1895), reprinted in *Skrifter* [Writings] (Stockholm: Bonnier, 1914), vol. 1, p. 282.

Oberstdorf has been singled out: Julia Boyd and Angelika Patel, *A Village in the Third Reich: How Ordinary Lives Were Transformed By the Rise of Fascism* (London: Elliott and Thompson, 2022).

Expeditions at end 1930s: Christopher Hale, *Himmler's Crusade: The Nazi Expedition to Find the Origins of the Aryan Race* (Hoboken, NJ: John Wiley and Sons, 2003).

Persistent myth of Hitler's bases in Antarctica: Ladislas Szabo, *Nazi Antarctic Exploration: Hitler's Escape to South America and Secret Bases in Antarctica* (Cheltenham: Reason Publishing, 2022). The myth that Hitler had bases in Antarctica is utterly absurd and lacks any semblance of credibility, but has nevertheless been thoroughly and meritoriously refuted in detail by C. Summerhayes and P. Beeching in 'Hitler's Antarctic base: The myth and the reality', *Polar Record*, 43:1 (2007), pp. 1–21; J. Whitfield, 'Did Hitler have a base in the Antarctic?', *Nature*, 30 March 2007, https://www.nature.com/news/2007/070326/full/news070326-14.html; U. Rack, 'The Third Reich in Antarctica: The German Antarctic Expedition 1938–1939', *Polar Journal* 3:2 (2013), pp. 470–2.

Camp Century: Kristian H. Nielsen and Henry Nielsen, *Camp Century: The Untold Story of America's Secret Arctic Military Base Under the Greenland Ice* (New York: Columbia University Press, 2021); W. Colgan et al., 'The abandoned ice sheet base at Camp Century, Greenland, in a warming climate', *Geophysical Research Letters*, 43:15 (2016), pp. 8091–6; B. Vandecrux et al., 'Firn evolution at Camp Century, Greenland: 1966–2100', *Frontiers in Earth Sciences*, 9 (2021), p. 578978.

British Empire made snow cult object: Tom Simpson, 'Imperial slippages: Encountering and knowing ice in and beyond colonial India', in Klaus Dodds and Sverker Sörlin (eds), *Ice Humanities: Living, Working, and Thinking in a Melting World* (Manchester University Press, 2022), pp. 205–27.

Australia's continent of snow: Brigid Hains, *The Ice and the Inland: Mawson, Flynn, and the Myth of the Frontier* (Melbourne: Melbourne University Press, 2002).

Norwegian gold prospectors: Jacob Vaage, *Skienes Verden* [Ski world] (Oslo: Hjemmets Forlag, 1979), p. 116.

SPEED AND HARMONY

Modern man climbing: Peter H. Hansen, *The Summits of Modern Man:*

Mountaineering after the Enlightenment (Cambridge, MA: Harvard University Press, 2013).

Landscape of skiing: Andrew Denning, *Skiing into Modernity: A Cultural and Environmental History* (Oakland: University of California Press, 2015), pp. 58–68.

'Industrialized nature': Paul R. Josephson, *Industrialized Nature: Brute Force Technology and the Transformation of the Natural World* (Washington DC: Island Press, 2002).

Sportification: Allen Guttmann, *From Ritual to Record: The Nature of Modern Sports* (New York: Columbia University Press, 1978).

World Cups for snow sports: Serge Lang and Patrick Lang, *La coupe du monde de ski alpin* (Grenoble: Éditions Jacques Glénat, 1986); Sébastien Cala, 'The Alpine Ski World Cup: A "game changer" for the International Ski Federation (1967–1975)?', *Sport in History*, 44:1 (2024), pp. 78–94.

DEATH IN THE SNOW

According to environmental historian of Russia, Andy Bruno: A. Bruno, 'Tumbling snow: vulnerability to avalanches in the Soviet North', *Environmental History*, 18:4 (2013), pp. 683–709. Bruno's article is the primary source of information about the avalanches and mining in the Khibiny area.

Railroad avalanche at Rogers Pass: Diana L. Di Stefano, *Encounters in Avalanche Country: A History of Survival in the Mountain West, 1820–1920* (Seattle: University of Washington Press, 2013).

Prioritization of production plans: Paul R. Josephson, *The Conquest of the Russian Arctic* (Cambridge: Harvard University Press, 2014); John McCannon, *Red Arctic: Polar Exploration and the Myth of the North in the Soviet Union, 1932–1939* (Oxford University Press, 1998).

Fersman's efforts inspired by Switzerland: Sergey A. Sokratov, 'The Russian contribution to snow science', *ICE* (2013), pp. 4–9; Christophe Ancey et al., 'Some notes on the history of snow and avalanche research in Europe, Asia and America', *ICE*, 139:3 (2005), pp. 3–11.

Preventive blasting: H. Gubler, 'Artificial release of avalanches by explosives', *Journal of Glaciology*, 19:81 (1977), pp. 419–29.

Avalanches in the First World War: S. Morosini, '"Following in the footsteps of history": sixteen multimedia itineraries through the First World War sites in the Stelvio National Park and Adamello Park (Italy)', in D. Svensson et al. (eds), *Pathways: Exploring the Routes of a Movement Heritage* (Winwick, Cambridgeshire: White Horse Press, 2022), pp. 114–37.

SELECTED REFERENCES

Winter of terror, 1951: Christian Pfister, *Wetternachhersage: 500 Jahre Klimavariationen und Naturkatastrophen (1496–1995)* [Weather forecast: 500 years of climate variations and natural disasters (1496–1995)] (Bern: Haupt, 1999), pp. 258–60; Eidg. Institut für Schnee- und Lawinenforschung, *Schnee und Lawinen in den Schweizeralpen, Winter 1950/51*, vol. 15 (Davos Platz: Buchdruckerei Davos, 1952), p. 180, https://www.dora.lib4ri.ch/wsl/islandora/object/wsl%3A17269

Switzerland focus on avalanche research: Christian Pfister (ed.), *Am Tag danach: Zur Bewältigung von Naturkatastrophen in der Schweiz 1500–2000* [The day after: coping with natural disasters in Switzerland 1500–2000] (Bern: Haupt Verlag, 2002).

Cost of avalanche barriers: Gerald Seligman, *Snow Structure and Ski Fields: Being an Account of Snow and Ice Forms Met With in Nature and a Study of Avalanches and Snowcraft* (London: MacMillan, 1936), p. 295.

Patriotic avalanche control: D. Achermann, 'Snow and avalanche research as patriotic duty? The institutionalization of a scientific discipline in Switzerland', in Julia Herzberg et al. (eds), *Ice and Snow in the Cold War: Histories of Extreme Climatic Environments* (New York: Berghahn Books, 2018), pp. 49–68.

In other parts of the world, too: 'Statusrapport avseende etableringen av lavinprognoser för svenska fjällen 2016' [Status report concerning the establishment of avalanche forecasts in the Swedish mountains], Naturvårdsverket [Swedish Environmental Protection Agency], 2016, https://www.diva-portal.org/smash/get/diva2:956069/FULLTEXT01.pdf

Bilgeri was multi-talented: G. Bilgeri, 'Alpin skidteknik' [Alpine ski technique], in *På skidor* [On skis] (1926), pp. 281–6.

Royal interest in ski sports: S. Sörlin, 'En nasjon krysser sitt spor: Konger, helgener og "banal Nasjonalisme" i den evige snöens rike' [A nation crosses its path: Kings, saints and 'banal nationalism' in the realm of eternal snow], *Samtiden* [Contemporary life] 122:1 (2011) , pp. 66–83.

Skiers were like frontline soldiers: K. Sandell and S. Sörlin, 'Naturen som fostrare: Friluftsliv och ideologi i svenskt 1900-tal' [Nature as instructor: Outdoor life and ideology in 20th-century Sweden], *Historisk tidskrift* [Historical journal] 114:1 (1994), pp. 4–43.

Triggering an avalanche with your voice: B. Reuter and J. Schweizer, 'Avalanche triggering by sound: Myth and truth', in *International Snow Science Workshop* (Davos: Swiss Federal Institute for Forest, Snow and Landscape Research, 2009), pp. 330–3.

Suffocated to death: Seligman, *Snow Structure*, p. 481.

METAMORPHOSES

Castorp's ski journey: Thomas Mann, *The Magic Mountain* (1924), tr. H. T. Lowe-Porter (London: Secker & Warburg, 3rd edn, 1961), pp. 472, 475–8, 480, 497. S. Sasse, 'Schneesturm im Gebirge: das Hans Castorp-Syndrom' [Snowstorm in the mountains: the Hans Castorp Syndrome], in G. Frölicher et al., *Dieser Mont Blanc verdeckt doch die ganze Aussicht!: Der literarische Blick auf Alpen, Tatra und Kaukasus* [This Mont Blanc obscures the whole view!: The literary view of the Alps, Tatra and Caucasus] (Norderstedt: Edition Schublade, 2016), pp. 239–54.

New model on snow and security: In 2018 UNESCO accepted 'Avalanche Risk Management' in Switzerland and Austria on to the List of the Intangible Cultural Heritage of Humanity, at https://ich.unesco.org/en/RL/avalanche-risk-management-01380

A major shift: D. Achermann, 'Snow and avalanche research as patriotic duty? The institutionalization of a scientific discipline in Switzerland', in Julia Herzberg et al. (eds), *Ice and Snow in the Cold War: Histories of Extreme Climatic Environments* (New York: Berghahn Books, 2018), pp. 55–9; R. Haefeli, 'The development of snow and glacier research in Switzerland', *Journal of Glaciology*, 1:4 (1948), pp. 192–201, https://www.cambridge.org/core/services/aop-cambridge-core/content/view/A0D1398F875CA522405260DF6F0E7BB9/S002214300000808Xa.pdf/div-class-title-the-development-of-snow-and-glacier-research-in-switzerland-div.pdf

'No matter what the circumstance': Commission for Snow and Avalanche Research (EAR) protocol, 11 December 1931, described in Achermann, 'Snow and avalanche research', in Herzberg et al., *Ice and Snow in the Cold War*, op. cit., pp. 53, 62.

Ahlmann, programmatic article: H. W. Ahlmann, 'Polarforskningens värde och berättigande' [The value and legitimacy of polar research], *Ord och bild* [Word and image] 41 (1932), pp. 195–207.

Snow – transformation: Durs Grünbein, *Om snön eller Descartes i Tyskland* [About the Snow or Descartes in Germany] (2002), translated from German to Swedish by Ulrika Wallenström (Stockholm: Ersatz, 2018). No earlier English translation is available, see https://www.suhrkamp.de/rights/book/durs-gruenbein-on-snow-or-descartes-in-germany-fr-9783518414552

Bader goes to the USA: 'Henri Bader (1907–1998)', *ICE*, 120:2 (1999), pp. 20–22.

Growing collaboration: Henri Bader and Daisuke Kuroiwa, *The Physics and Mechanics of Snow as a Material* (Hanover, New Hampshire: US Army Cold Regions Research and Engineering Laboratory, 1962).

USA enlists capabilities of other countries: John Krige, *American Hegemony and the Postwar Reconstruction of Science in Europe* (Cambridge, MA: MIT Press, 2006).

Word eventually became: 'environment': Paul Warde et al., *The Environment: A History of the Idea* (Baltimore, Maryland: Johns Hopkins University Press, 2018).

The cold was strategic: Stephen Bocking and Daniel Heidt (eds), *Cold Science: Environmental Knowledge in the North American Arctic during the Cold War* (New York: Routledge, 2019).

AFTERWORD: THE ANGELS' CONVERSATION

The Angel of History: Walter Benjamin, 'Theses on the philosophy of history', *Illuminations*, ed. H. Arendt, tr. Harry Zohn (New York: Schocken Books, 1969).

Acknowledgements

In my long life with snow, I have met too many people to be able to thank everyone who has meant something to this book. Some of them are now gliding forward on purer snow-fields. Others remain among us, still on Earth. Regardless of where they belong, some of them appear in the book, in text or references, which should be understood as an implicit 'thank you'.

It feels important to thank by name those I asked for special services as the project neared completion. Friends, colleagues, specialists from whom I have been able to get advice and/or who have read and commented on all or parts of the script: Andy Bruno, Anders Cullhed, Kjell Danell, Klaus Dodds, Anna Froster, Erik Isberg, Harald Dag Jølle, Isak Lidström, Lucas Mueller, Ninis Rosqvist, Bernhard Schirg, Daniel Svensson, Floris Winckel, Nina Wormbs. Thank you very much! Your comments and suggestions have been invaluable. Denise Hagströmer made sure that I ended up in the right place among the paintings and texts at the National Museum in Oslo. Peder Anker, for a joint essay a long time ago, collected reviews of Ukichiro Nakaya's *Snow Crystals*, from which I have quoted a couple of passages. My daughter Julia Olofsson Sörlin did a great job of arranging and checking the raw material for the references, which, despite being a selection, turned out to be far more numerous than I had thought they would be. Julia's younger sister, Lydia Bandolin

ACKNOWLEDGEMENTS

Sörlin, has helped primarily through conversations out in the snow, often on skis. I was able to find some time for research and writing during my stay as a fellow at the Swedish Collegium for Advanced Study, SCAS, at Uppsala University in 2022–3.

At my Swedish publisher, thanks are owed to editor Mattias Pettersson, publisher Johan Wirfält and designer Lukas Möllersten, who also created the design for the British edition. Thanks to Elizabeth de Noma for the English translation and to Alex Newby for her work on the British text. And to Matilda Berg and Ulrika Bergwall in Sweden and to Hannah Winter and Millie Seaward in Britain for spearheading conversations with readers about snow and our common existential conditions. Because even a book about snow is ultimately a book about us humans.

I dedicate the book to my mother Gudrun, born in 1933, in gratitude. She used all her senses to open up the worlds of snow to me.

Sverker Sörlin
Åsele and Stockholm, July 2025

Sverker Sörlin is an author, historian and science communicator. He is currently Professor of Environmental History at the KTH Royal Institute of Technology in Stockholm, and during his career he has worked with universities and institutes in Berkeley, Princeton, Cambridge, Vancouver, Cape Town, Oslo, among others. He is a defining voice in Swedish public affairs and has served as an advisor to the Swedish government on science and climate policies.

He has published more than forty books, including bestselling literary non-fiction, biographies, academic works, collections of journalism, and personal essays on topics ranging from climate change, the Anthropocene and Charles Darwin to gout, popular education and cross-country skiing. In 2004 he received the August Award for Non-fiction, Sweden's pre-eminent literary award, and in 2024 was awarded the Inge Jonsson Prize by The Nine Foundation for outstanding non-fiction.